Frame of the Universe

Frame of the Universe
A History of Physical Cosmology

Frank Durham
Robert D. Purrington

NEW YORK
Columbia University Press

Library of Congress Cataloging in Publication Data

Durham, Frank, 1935–
Frame of the universe.

Bibliography: p.
Includes index.
1. Cosmology—History. I. Purrington, Robert D.
II. Title.
QB981.D89 1983   523.1'09   82-12990
ISBN 978-0-231-05393-8 (pbk.)

Columbia University Press
New York   Chichester, West Sussex

Contents

Preface

This book is a history of scientific ideas written not by historians of science, but by practicing scientists. That we each bring to this work the perspective of a quantum physicist with a special understanding of and sensitivity to developments at the frontiers of physics during the last quarter century, coupled with a long-standing interest in the history of astronomy and cosmology, gives us an unusual advantage in attempting a work such as this. This advantage is especially obvious when dealing with the very modern topics, many of which are to be found only in the primary literature. There is, of course, a concomitant danger, which is that our understanding of the processes and methodology of history will be less fully developed, and that our reading of the data of history will be found wanting. We can see no escape from this dilemma, short of keeping our ideas to ourselves. The history of cosmological thought developed here is as we have come to understand it in over ten years of reading, and of teaching the subject to university undergraduates. Most of the ideas presented here (at least in the first eleven chapters) were learned from professional historians of science. Where we have made our own judgments and interpretations, mainly in the last half of the work, we have tried to make that clear, and we invite the reader to pursue the questions raised in these discussions in the vast literature of the history of science and to consult the original sources. At every point in the narrative, there will be experts whose ideas should be explored by the reader interested in a deeper and more detailed study of the particular period or historical figure—e.g., a Stillman Drake on Galileo or a Cohen or Westfall on Newton. The bibliography is sufficiently detailed to provide an introduction to the relevant literature. The primary journals in the field, such as *Isis, Archive for History of the Exact Sciences, Journal for the History of Astronomy,* and so on, are not often mentioned in the bibliography, but represent the battleground on which many of the ideas discussed here are developed and tested and the interested reader with access to a good library should not forget them.

An additional danger is that of focusing on a completed idea as opposed to the way in which that idea evolved. Scientists are notoriously guilty of this ahistorical view, but it is not universal, and we hope to have avoided it. We have also endeavored to keep our own philosophical prejudices as much as possible in the background. We have separately cut our philosophical teeth on Thomas Kuhn and Karl Popper, a fact that may at one point or another be evident from our discussion. We have tried to exclude these considerations, to avoid seeing the history of cosmology from the point of view of any particular philosophy of scientific change, but no doubt we have not always succeeded.

Our bias is that of physics and astronomy, disciplines which claim cosmology as a subfield. As we trace the development of cosmological themes we shall encounter many of the great ideas of western civilization. Some of these ideas will seem to have little to do with physics or astronomy. Indeed, we shall find a considerable scientific content in the records of ancient cultures, in their myth-making and in the prehistoric remains which contain all that is known of some primitive peoples. Yet in tracing the history of cosmology, the mutually imposed separation of the physical sciences from other intellectual disciplines cannot long be maintained.

We argue that man fully understands himself only if he understands the universe of which he is a part, that we are in a fundamental way inseparable from the universe that spawned and nurtures us. It may be that there is no single way of perceiving the universe, and certainly not everyone would agree that the astrophysicist's description to which we shall come in the end is exhaustive, that it encompasses all that can be said of the nature and origin of the universe. It is, nevertheless, one important piece of this greatest puzzle, one aspect of the truth about the world. And in this case, at least, the way is clear and the methodology agreed upon, even if one may despair of achieving complete understanding even on this level. There are new discoveries to be made, and the darkness which bounds our existence is pushed back imperceptibly with each one. Scientists have sometimes been criticized for equating scientific progress with the progress of civilization, or for seeming to argue that science is the supreme example of man's intellectual growth. In some cases the criticism is telling, and yet the directed course of scientific development, while not without its retreats and blind alleys, leads inexorably to a deeper understanding of the natural world, in a way that has no parallel in other areas of human affairs. To see a sort of perfectibility in science is not to elevate scientists above the rest of mankind; rather it is to recognize the efficacy of the manner of explanation. This is itself a consequence of the limit science has imposed upon itself, to describe only what is accessible to measurement.

The perspective for which we search is in a sense impossible of achievement, since it is dynamic and changing. It is not enough to describe the truth as we now see it; we must understand how that perception has evolved, how it has been shaped by the culture. So while we shall conclude this work by establishing the late-twentieth-century worldview, we shall begin with the earliest awareness of the universe into which we have been born. Through an understanding of how we came to see the universe as we do we may better see our present worldview for what it is: not a final statement, but a conditional and temporary consensus. We may hope, as well, to gain some insight into the future evolution of ideas about the universe and our place in it.

We should like to acknowledge discussions with many students over the past ten years, discussions which helped shape our understanding of the history of cosmology and its place in the intellectual history of man. Especially important has been the influence of Dr. Alan Goodman, who has taught some of this material with us and who has brought his own profound insights to the classroom and to informal discussions of the material. Discussions with George Rosensteel, McKin Malville, and Ernan McMullin are gratefully acknowledged.

Frank Durham
Robert D. Purrington
October 1982

Frame of the Universe

CHAPTER 1

Overview

For the artist and for the scientist there is a special problem and a special hope, for in their extraordinarily different ways, in their lives that have increasingly divergent character, there is still a sensed bond, a sensed analogy. Both the man of science and the man of art live always at the edge of mystery, surrounded by it; both always as the measure of their creation, have had to do with the harmonization of what is new with what is familiar, with the balance between novelty and synthesis, with the struggle to make partial order in total chaos. They can, in their work and in their lives, help themselves, help one another, and help all men. They can make the paths that connect the villages of arts and sciences with each other and with the world at large the multiple, varied, precious bonds of a true and worldwide community.

This cannot be an easy life. We shall have a rugged time of it to keep our minds open and to keep them deep, to keep our sense of beauty and our ability to make it, and our occasional ability to see it in places remote and strange and unfamiliar; we shall have a rugged time of it, all of us, in keeping these gardens in our villages, in keeping open the manifold, intricate, casual paths, to keep these flourishing in a great, open, windy world; but this, as I see it, is the condition of man.

J. Robert Oppenheimer

One of man's great preoccupations has always been with the nature of the universe: what it is made up of, how it came into being, how it will end. One has only to visit a major museum where there will be found Babylonian boundary stones with their representations of the Sun, Moon, Venus, gods representing the forces of nature, or perhaps Babylonian cylinder seals, on which are portrayed scenes of creation and strife among the gods of nature and the sky. In a collection of Egyptian artifacts one might find a drawing or painting of the Egyptian cosmos, with the sky, the air, and the earth personified as deities with specific roles, or encounter a papyrus manuscript containing a prayer to the sun-god. A visit to Stonehenge in southern England or to the ruin of

Chichén Itzá in Yucatan, each a concrete example of horizon astronomy, should remove any doubt about the enormous interest of primitive cultures in the skies and, more importantly, in the nature and origin of the universe.

As we shall soon see, these issues have troubled man from time immemorial. They are expressed in the preliterate cosmological myths that still survive among certain primitive cultures; they lie at the source of the mythical cosmologies found in such epics as the early Babylonian creation epic, the *Enuma Elish,* and form the basis for the Hebrew creation myth found in Genesis. For millennia, Western man gave no other expression of his speculations about creation than through myth.

Eventually, by the middle of the first millennium B.C. in the early Greek cities of Ionia, rational speculation about the universe began. The tools of imagination and fancy were put aside, slowly at first, as man set about systematically to try to understand the universe and its origins. This crucial understanding that the universe is rational rather than contingent and under the sway of arbitrary or capricious gods did not occur abruptly, but it represents one of the great watersheds in the history of human thought.

The cosmological ideas of the early Milesians might justifiably be described as naive; certainly their understanding of the natural world was inadequate. Yet the willingness to generalize from the concrete and particular to the abstract stands out clearly, in contrast to the earlier efforts of the Near Eastern civilizations.[1] By 500 B.C. or shortly after, we find Greek thinkers, the earliest philosophers, not only speculating about the origin of the world but also committing their ideas to writing. Such men as Anaximander, Parmenides, and Pythagoras set about systematically to try to understand the universe and its origins. Within a century Aristotle had provided a synthesis that would not be surpassed for sixteen centuries.

Yet the flowering of the Greek intellect was destined to be short lived. Under Roman domination the practical arts and astrology flourished. The Hellenistic philosophies of Stoicism and Epicureanism had only the slightest use for scientific inquiry. Predictive and computational astronomy began to take the place of rational speculation about the workings of the heavens and of planetary dynamics. The Ptolemaic planetary system, that greatest achievement of Hellenistic astronomy, was a product of a pragmatic Greek science past its creative prime. Yet no other concept of the universe so long endured unchallenged. Finally, well into the sixteenth century, Copernicus took the revolutionary step that provided the basis for the sun-centered cosmology Kepler and Galileo were to establish.

The interim, the ten centuries of medieval Europe, are not without importance in this story. Of particular interest is the manner in which the ancient Greek traditions and knowledge were finally transmitted to the reawakening Christian West. During this period as well, the foundations of mechanics were being laid in England and France. Yet in the last analysis, it was the planetary dynamics of Copernicus and Kepler and the mechanics and astronomy of Galileo that fueled the revolution that set the world aflame with the spirit of scientific rationalism. Dominating this episode is the giant figure of Newton. The seventeenth century ended with the earth adrift, and the hand symbolically at the tiller was that of Isaac Newton. His universe is a mechanism governed by commonsensical laws and undergirded by a single, mysterious agent: universal gravitation.

In the published work of Galileo and Newton we see "why?" replaced by the manageable "how?" of physics. Thus circumscribed, cosmology has nevertheless continued to fascinate practitioner and casual observer alike. The very breadth and subtlety of the descriptions provided in the twentieth century by observational astronomy and modern physics have sufficed; the means have become the end. Still, the possibility of a complete answer to the riddle of the universe remains a dream.

Astronomy in the eighteenth and nineteenth centuries produced few theoretical or analytical triumphs; the achievements were mainly observational. New instruments, new techniques led to the discovery of new and puzzling phenomena and provided the raw material for the theoretical developments in twentieth-century astrophysics and cosmology. Coming to understand the "island universes," other galaxies, must rank high in a list of this century's achievements. The discovery of the universal recession of these objects opened the way for modern cosmology. Einstein's general theory of relativity made possible an understanding of these large-scale features of the universe and, with the special theory, overthrew the Newtonian concepts of space and time. The development of quantum mechanics not only thoroughly shattered Newtonian materialism but also made possible an understanding of energy generation in stars, so that by mid-twentieth century the structure and evolution of the stars could be described. The great questions about the origin of the universe remain incompletely answered, but the vast panorama offered by modern astronomy overwhelms the imagination with exhilarating implications for cosmology.

What new truths remain to be uncovered, what great and perplexing discoveries remain to be made, none can predict. Modern science can tell us about the future evolution of the sun, it can venture astronomical explanation of the ice ages and of biological evolution, it can provide a

sound basis for statements about the future of the universe, whether
from the direction of general relativity or the second law of thermodynamics. It cannot, however, tell man how to handle the weapons of
destruction he has fashioned, with which he seems capable of destroying his civilization, or how to solve the problem of climatic changes
attendant upon atmospheric pollution, which may radically alter life on
the planet. In the end, we are left with more questions than answers.
Nobel Laureate Isidor Rabi has offered this view of our search for understanding.

Scientific understanding . . . is an essential step to our finding a home for
ourselves in the universe. Through understanding the universe, we become at
home in it. In a certain sense we have made this universe out of human concepts
and human discoveries. It ceases to be a lonely place, because we can to some
extent actually navigate in it.[2]

Unconscious Newtonianism

Inevitably there will be those who will not travel the whole road we
have just described, who will retreat from the scientific world view at
the point at which it becomes, to them, oppressively quantitative, ex-
cessively complex. Though they may be excited by the prospect of
learning about primitive protoscience and find the complex personalities of Kepler and Galileo fascinating, they begin to get uncomfortable
at about the eighteenth century, and before the first neutron star is
sighted they are gone—at least in spirit. They have no need for a vision
of the modern universe; a glimpse of its complexity is enough. They
will settle for the "natural" world. We argue that an acquaintance with
the physical laws in their historical context does not sustain an accusa-
tion of oppressiveness or of coldness, but this outlook on the problem
of the relation of science to the rest of knowledge is a personal one.
Many people see the unfolding of the universe according to physical
law as implying the reduction of life to an irrelevancy. Yet it is also
possible to regard the achievements of Galileo, Newton, and Einstein
as a reflection of every person's need "to struggle for a greater, more
comprehensive version of himself," as Loren Eiseley has put it.[3]
    In any event a retreat from physics to the "natural" world of trees
and grass is bound to be frustrated. This is true, at least in Western
countries, because the language itself is permeated with scientific concepts. The person who relies on the simplest, most colloquial descriptions of ordinary objects and everyday events may think he has evaded
mathematics and physics, and he is largely right insofar as twentieth-

century physics is concerned. But his world view nevertheless is not primitive. It is certain to be in most respects Newtonian, by which we denote the cause-and-effect materialism that came to prominence in the seventeenth and eighteenth centuries and that was retained in large measure because of the success of Isaac Newton's physical theories.

In this century the language of Newtonian physics has been shown not to provide a complete or even adequate description of physical phenomena, yet its effect lingers. This is true, not only in common usage (in questions such as "What time is it?"), but also in our social institutions (as when we speak of "natural" rights). In the Newtonian world view, space, time, and matter are the elements of the physical world and are to most of us so obvious that they hardly require comment. Yet the world is not obvious or simple; space and time are not the primitive, commonsense entities we usually take them to be. It is, moreover, by no means clear that any description of the world can be regarded as unique. What then is the average person to do in the face of this complexity? In the modern world, refusing to address the question means uncritically accepting someone else's world view, whether scientific or not.

Recent years have seen the idea that there exist in Western civilization two disjunct cultures with irreconcilable differences. On the one hand there is supposed to be a scientific-technological culture, analytical and quantitative, eschewing the transcendent element of life. This is contrasted to the nonscientific culture, exemplified by the liberal and fine arts, which view nature subjectively, seeking a unified picture of nature rather than an analytical one and rejecting the mechanistic world of science.

This dichotomy is less discussed now than it was a decade ago, when advocates of an antitechnological counterculture were numerous. The present quiescent state can be traced probably to the end of the Vietnam War and to the failure of the alternative life styles proposed by writers such as Theodore Roszak, rather than to any contentment with the technological society. Surely the widespread interest in anti-intellectual pursuits, such as astrology and the occult, is symptomatic of a later stage of malaise. It is as if following a logical argument—even an argument for irrationality—has become too taxing for many.

By contrast the scientist can be seen as sharing with the painter or writer a need for beauty and for a comprehensive sense of his place in the world. The scientist does nonetheless represent a particular point of view about how the world is organized: that it is to be understood in rational terms, that however multileveled the nature of truth and of human experience, there is a level of truth for which the methods of science are uniquely suited. How man integrates this understanding into his total perception of himself and the world is another matter, best dealt with by the philosopher, the mystic, the theologian, the poet. The poet's vision of the natural world, neither more nor less valid than that of the scientist, is assuredly different:

> Driving at night I feel the Milky Way
> streaming above me like the graph of a cry[4]
> (Adrienne Rich)

or from e.e. cummings:

> While you and I have lips and voices which
> are for kissing and to sing with
> Who cares if some one eyed son of a bitch
> invents an instrument to measure Spring with?[5]

Yet it was Albert Einstein who said, "The most beautiful thing we can experience is the mysterious. It is the source of all the true art and science. He who can no longer pause to wonder and stand rapt in awe is as good as dead: his eyes are closed."

In a sense the Newtonian world view is now archaic. Thoroughly engrained in our existence, it is nonetheless no longer tenable. As relativity shattered Newtonian ideas about space and time, so the quantum theory undermined the edifice of deterministic materialism that was erected on the foundation of Newtonian physics. Collectively modern physics demands an entirely new epistemology, a new view of what measurement is. Lest you be troubled that we give this second great revolution of twentieth-century physics, quantum mechanics, short shrift, we remind you that this is a history of cosmological thought, to which the quantum theory has only begun to contribute in a profound way. A detailed description of the early moments of the universe, or of those extraordinary pathological objects, the white dwarf and neutron stars, would require the full apparatus of quantum physics. To do this, and to trace the evolution of these ideas, would expand this narrative beyond reasonable bounds and would only slightly elucidate our understanding of how we come to see the universe as we do. Thus, where the quantum theory does intrude, we endeavor to make its importance and ramifications clear, without attempting a full exposition of these ideas.

As we proceed, chronologically, to unfold the story of mankind's changing perception of the universe, we shall encounter one transformation in scientific thought after another. Some will seem clearly to deserve the term *revolution;* others will not. In some cases, perhaps, the metaphor will be totally inappropriate. We will for the most part try to be descriptive rather than analytical, to say "this is what seems to have taken place," rather than to say why or to generalize to a theory of scientific change. But be aware that some of the better minds of our time have devoted much of their lives to these questions.[6] The issues are vital and are hotly debated. We mention them only in passing.

CHAPTER 2

Earliest Awareness

By the middle of the twentieth century, nuclear dating techniques, especially the dating of meteorites, had established the age of the solar system at slightly more than 4.5 billion years. This provided a lower limit on the age of the universe, which is obviously more difficult to obtain, and radioactive dating combined with astronomical results based on the theory of stellar evolution now give a probable age for the universe of more than 15 billion years. This number is quite recent—as late as 1950 the age of the earth seemed greater than the derived age of the universe. Our goal is, of course, to trace the evolution of cosmologic ideas, including the question of the age of the universe, from early man to the present. Our attention will be drawn especially to opinions about how the universe came into being.

Early Man and the Heavens

Unfortunately for the historian, mankind's awareness of the regularities of the sun, moon, and stars was already highly developed by the dawn of recorded history, nearly 6,000 years ago. For although the history of mankind goes back more than 2 million years, the fossil evidence on which knowledge of our origins is based permits only the most tentative speculation about our mental development. The situation is not hopeless, however, and enormous progress has been made through a multidisciplinary attack using the techniques of anthropology, linguistics, astronomy, and comparative mythology on this complex problem of tracing preliterate knowledge. Nevertheless, many of the fundamental questions about the development of man's relationship to nature remain unanswered.

What do we know of his early wondering at the stars and their motions or at the cyclic alternations of the sun and moon? When did we come to realize how deeply regular these phenomena are, and when did we begin to use them as a measure of the passage of time? This discovery—that certain phenomena are repetitive, their recurrence predictable, that they can be used to partition time—must have had an importance for the intellectual development of mankind as great as the wheel or fire were for technological advancement. Yet of this "skylore" there frequently remains no concrete evidence. What evidence there is we shall examine in the next chapter, but most often our knowledge of mankind's earliest interest in the skies comes from oral traditions that survived from the prehistoric period. These provide only the faintest hint at how stone-age man saw the world around him, and the heavens in particular, but we know that by the time he began to be able to commit his ideas to writing, his interest in the stars and planets and the celestial regularities had already had a long history. For the moment we turn to the expression of his interest in the heavens as embodied in myths, and in particular his description of how the universe came into being.

Myth and Pre-Science

Although early man possessed a mental capacity little different from our own, he was faced with a physical and psychic world of which he had little understanding. The erratic, unpredictable forces of nature were awesome and fearsome. Eventually the orderly, repetitive phenomena, mostly celestial, were sorted out of the chaos of experience; on these are based our timekeeping. These two features, the terrifying and mysterious nature of life and the contrasting order of the heavens, were important in molding the myths that early man began to weave out of the stuff of existence. Far from being mere legend or saga, these myths express the universal elements of human experience in terms real to the people who preserved and lived them. The most accessible examples of myth in this high sense are found in the early books of the Old Testament. The truth of myth is of this world but of the next as well; "it is a form of poetry which transcends poetry in that it proclaims a truth; a form of reasoning which transcends reasoning in that it wants to bring about the truth it proclaims; a form of action, of ritual behavior, which does not find its fulfillment in the act, which must proclaim and elaborate a poetic form of truth."[1]

For many of us the myths are dead. Try as we may, we cannot breathe life into them. Our relation to nature is totally unlike that of the mythmakers, which was personal. This relationship has been described as "I-thou," endowing the inanimate world with the attributes of a living being. Every feature of the environment had to be contended with; a flooded river might need to be cajoled or pacified, a falling rock to be outwitted or cursed. The natural world was a battleground of conflicting powers: sun, wind, death—powers that have become for us mere things only, or impersonal events. While we look for single, unifying causes for typical phenomena (gravity, for example, "explains" every falling object), there were for primitive man no underlying material causes. Experience was too immediate for explanation, and there was no effort to simplify it. The essence of myth is the use of many-sided images to express the various aspects of a single phenomenon. Thus, the Frankforts wrote,

The Egyptians in the earliest time recognized Horus, a god of heaven, as their main deity. He was imagined as a gigantic falcon hovering over the earth with outstretched wings, the coloured clouds of sunset and sunrise being his speckled breast and the sun and moon in his eyes. Yet this god could also be viewed as a sun-god, since the sun, the most powerful thing in the sky, was naturally considered a manifestation of the god and thus confronted man with the same divine presence which he adored in the falcon spreading its wings over the earth.[2]

The Egyptians and Mesopotamians, whose myths provided the first Western cosmologies, lived with simultaneous, incompatible representations of nature.[3] Modern man has fled from ambiguity and from immediacy in nature:

                    the lights in the sky are stars
We think they do not see
                    we think also
The trees do not know nor the leaves of the grasses
                                        hear us[4]

At an early date astronomical images and symbols began to play an important role in myth. This was ensured by the regular, orderly nature of certain celestial phenomena, which quickly impressed themselves on the mind of early man. Gods became identified with important stars, planets, or constellations, and interest in astronomical cycles gave rise to complex images reflecting psychic conflicts, for example, the almost universal myth of the resurrected hero, traced by Frazer to the regular death and rebirth of the moon. In fact the universality of many of the images contained in early myth is an intriguing aspect of these myths. The origin of this universality has challenged comparative mythologists from the dawn of that science. Is it due to the universal and shared nature of human experience, the mysteries and terror of birth and death, of the forces of nature, or is it innate? Are our myths somehow a product of innate structures of our nervous systems, which govern our behavior, as suggested by sociobiology? Or is there yet another explanation, akin to these, but originating in man's common experience of the heavens?

Strong proponents of an astronomical basis for much of the content of archaic myth are de Santillana and von Dechend. They argue, in *Hamlet's Mill,* that no other experience of mankind is as universal as that of the heavens; and the Frankforts write that "the imagery of myth is . . . nothing less than a carefully chosen cloak for abstract thought representing the form in which experience has become conscious."[5] This view is controversial, but many myths are indeed global, and there is reason to believe that the events described occurred in the sky and not merely on the earth. In *Hamlet's Mill,* von Dechend and de Santillana interpret the puzzling biblical story of Samson (Judges XV) in astronomical terms:

15. And he found a new jawbone of an ass, and put forth his hand and took it, and slew a thousand men therewith.
16. And Samson said, with the jawbone of an ass, heaps upon heaps, with the jaw of an ass have I slain a thousand men.
17. And it came to pass, when he had made an end of speaking, that he cast away the jawbone out of his hand, and called that place Ramathlehi.

What is argued is that the jaw is not an ordinary bone; rather it is in heaven. To the Babylonians it was the Hyades Cluster, the jaw of Taurus the bull. Specifically,

In the Babylonian creation epic, which antedates Samson, Marduk uses the Hyades as a boomerang-like weapon to destroy the brook of heavenly monsters. The whole story takes place among the gods. . . . If one brings Samson—the biblical Shimshon—back to earth, he becomes a preposterous character, or rather not character at all, except for his manic violence and his sudden passions.[6]

One particularly striking phenomenon of the heavens probably contributes to the imagery of different global myths. It is the *precession of the equinoxes.* "Precession" is the motion in space of the earth's axis, the imaginary line through the north and south poles. The earth turns once around its axis daily, so that an observer on the earth sees the stars

Figure 2.1 Time exposure of stars "turning" about the north celestial pole. Photo by permission of Lick Observatory.

carried around the sky. (fig. 2.1) A star that happens to be near either the celestial north pole (fig. 2.2) or the celestial south pole will move hardly at all. Now, like a spinning top, the earth gradually changes its direction or rotation. This is the phenomenon of precession, which carries the earth's axis very slowly around a cone of angle 47° (fig. 2.3). Every 26,000 years the north pole returns to point at the same star.[7] The axis now points toward the star Polaris (within 1°), but in 14,000 years the pole star will be Vega, and Polaris will swing in a wide circle daily. For the early Egyptians the pole star was α Draconis, the brightest star in the constellation Draco.

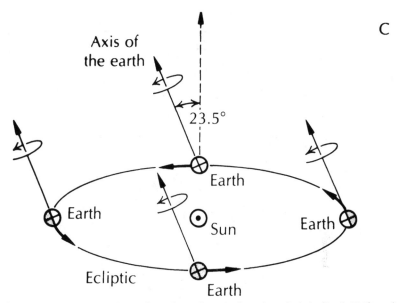

Figure 2.2A As the earth revolves about the sun, its spin axis is inclined 23.5° to the perpendicular. The star toward which northward extension of the axis points is called the "pole star," currently Polaris. L. Oster, *Modern Astronomy* (San Francisco: Holden-Day, 1973).

The effect of precession is a gradual but profound shifting of the heavens, hardly noticeable over periods of a few hundreds of years but dramatically evident over thousands of years. It was not until the second century B.C. that precession was "discovered" as a quantitative phenomenon. Of course there is no explicit reference to "precession" in the earliest literature, but the phenomenon is, nonetheless, hidden there. Of this phenomenon there was awareness without understanding. Scattered throughout world literature are references to legends of "the unhinging of the mill," to the drowning of various constellations or their personifications. Precession changes the pole star, "unhinging the mill" and casting constellations below the horizon. This is an awesome process, because it is not reversible; it undermines the regularity of the heavens. The Babylonian *Gilgamesh* epic says, in a passage which seems to refer to precession:

When I stood up from my seat and let the floor break in then the judgment of Earth and Heaven went out of joint. . . . The gods, which trembled, the stars of heaven—Their positions changed, and I did not bring them back.[8]

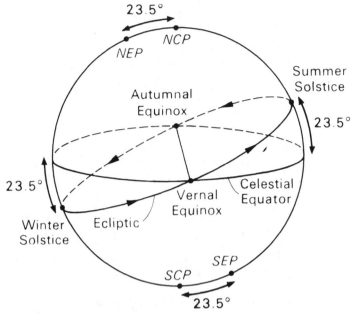

Figure 2.2B The "celestial sphere" is both a useful and historically important way of representing the sky as seen from the earth. The moon, sun, planets, and stars are all represented as though they all have the same distance from the earth, which they of course do not. The celestial sphere rotates once daily (because of the earth's rotation). The ecliptic is the apparent path of the sun on the celestial sphere during a year. The planets follow the ecliptic to a greater or lesser degree, depending on the inclination of their orbits. The twelve zodiacal constellations form a band which straddles the ecliptic. Stanley P. Wyatt and James B. Kaler, *Principles of Astronomy* (Boston: Allyn and Bacon, 1981).

A further element of global myth, the sense of the "ages of man," similarly has its origins in precession. The sun's path through the sky, called the ecliptic (fig. 2.2), carries it through the series of twelve constellations that we know as the zodiac. Because of precession, the sun reaches a given zodiacal constellation slightly earlier each year. This process is so slow that more than 2,000 years is required for a shift of a full constellation (26,000 years divided by 12 zodiacal constellations). Although the sun at the Northern hemisphere's vernal equinox has been in Pisces throughout the Christian era (thus, in part, the use by the early Christians of the symbol of the fish), the next constellation moving backward along the ecliptic, Aquarius, will rule the equinox after only several hundred more years. Those who believe that Aquarius will bring a new age are in tune with an element of archaic myth. This kind of association allowed the Roman poet Vergil to predict that a new era

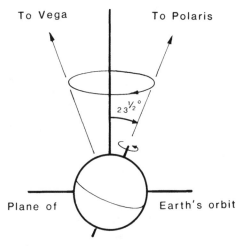

To Vega   To Polaris

$23\frac{1}{2}°$

Plane of   Earth's orbit

Figure 2.3 Precession of the earth's axis. The earth resembles a spinning top. Because the earth is not perfectly spherical, the moon (and sun) combine to cause the axis to turn slowly about the ecliptic pole (the direction perpendicular to the plane of the earth's orbit). This turning, called precession, carries the earth's axis around the ecliptic pole once every 26,000 years. As a consequence, the pole star is continually changing, as is the relationship between the seasons and the zodiacal constellations.

would begin when the sun entered Pisces, in what came to be called the first century Anno Domini. Very suggestive as well (see *Hamlet's Mill*) is the use by early Jews of the symbol of the Ram (Aries, which had preceded Pisces as the spring constellation), as against the pagan god symbolized by a golden calf (Taurus, the bull, an "obsolete" spring constellation). A dramatically symbolic passage in Chapter 12 of Revelation may also refer to precession. Finally, there is the possibility that Plato's Atlantis myth is a description of precession.[9]

This survey of early myth and its astronomical content only scratches the surface. Mircea Eliade and others have made profound studies of the general problem of myth, its origin and functioning, and its relation to celestial phenomena as well. Our focus being on the evolution of cosmological ideas as embodied in myth, we now turn to two of the earliest civilizations, those of Mesopotamia and Egypt, and to their cosmological myths.

The Cradle of Civilization

As early as 6000 B.C. the crucial transition to agriculture had occurred in the Fertile Crescent of Mesopotamia, and stable village communities had been established. No longer content to gather wild grains, these upper Neolithic farmers had begun to sow and reap on a systematic basis. With the changing climate of the postglacial era, marshes and swamps along the river valleys began to dry up, making their fertile soil available for cultivation. Thus by the opening of recorded history, a complex society already existed, not only in Mesopotamia, but also in the Nile Valley. The advent of village communities came somewhat later there than in the Tigris-Euphrates Valley, around 4500 B.C., and in the Indus Valley this stage was reached around 3000 B.C. The rapid development of agriculture led inevitably to the establishment of "urban" centers, city-states organized for defense and to satisfy other special needs such as the building of the dikes and canals for water control. The earliest Mesopotamian civilization centered in Sumer, on the Tigris-Euphrates Delta, near the head of the Persian Gulf, dominated by the cities of Ur, Nippur, Eridu, and Uruk. The food surpluses afforded by systematic agriculture made possible a leisure class from which the rulers or intermediaries to the gods came and provided a living for artisans and traders as well. Free of the need to secure life by heavy toil, the rulers or priests could turn their attention to problems of religion and social order. The rapidly developing civilization required accurate reckoning in order to calculate the time of the rising of the rivers or the time for reaping and sowing. A reliable calendar was thus needed. Navigation on land and sea also required knowledge of the motions of the sun and stars. Moreover, the vagaries of military fortune and uncertainties of farming required a knowledge of what the future held, in response to which astrology developed.

At about the same time civilization emerged in Mesopotamia, early Egyptian civilization had its own beginnings, concentrated along the wondrously fertile Nile Valley. Climatically Egypt was not unlike Mesopotamia. There was, as now, virtually no rainfall, and agriculture was possible only because of the regular flooding of the Nile River near the summer solstice. On the other hand, Egypt was virtually isolated from the external world, surrounded by desert and protected on the north by the sea, so that, unlike Mesopotamia, whose history was dominated by internal and external strife, Egypt had a relatively peaceful and stable history from about 3000 B.C. until the conquest by Alexander in 332 B.C. The relative isolation of the Egyptians resulted in a unique and somewhat static culture with a continuity unknown in Mesopotamia.

Writing developed independently in Egypt, where it took the form of hieroglyphics. Papyrus was used as a writing material, and because it could be rolled into scrolls many complete Egyptian texts have come down to us, while the clay tablets of Mesopotamians encouraged scattering or loss of parts of a text. The social structure that evolved in ancient Egypt differed crucially from its Mesopotamian counterpart. There were no true urban centers, and the existence of a divine king with absolute authority produced a strong central rule from an early date. During the third millennium B.C., in spite of the natural barriers separating the civilizations, Mesopotamian culture seems to have had a substantial influence on Egypt, particularly in the arts.

The relatively static character of early Egyptian history, coupled with the extreme regularity of rise and fall of the Nile and comparatively constant climate, free for the most part from the violence of nature, helped mold a reliable and comforting cosmology. Nature was orderly, governed by all-powerful gods, and the Pharaoh himself had divine powers. One consequence of this was a rather low interest in omen astrology in Egypt. In Mesopotamia, on the other hand, the social fabric was periodically rent by internal strife and external attack, while frightening storms brought unexpected floods in the great rivers to destroy the dikes and canals on which agriculture depended. The resulting Mesopotamian world view had a chaotic and unpredictable character, personified by warring gods of a demonic nature. Moreover, the local rulers were mere humans, only intermediaries to the gods whose favor might be withdrawn at any time. Thus the theme of chaos and conflict dominates the Mesopotamian epics and in particular the Babylonian Genesis, the *Enuma Elish*.

Babylonian Cosmology

The earliest Sumerian myths date from at least the third millennium B.C. but have preliterate origins that undoubtedly go back much further. The best known of these is the *Gilgamesh* epic, which deals at length with the problem of death and contains a description of the Great Flood. It is concerned with cosmology and cosmogony only incidentally. Heaven and earth were once united, and upon their separation, the heaven-god Anu carried off the heavens and the air-god Enlil carried off the earth. A much more specific and detailed picture of creation is to be found in the *Enuma Elish*[10] ("when above"), which had its origins in Akkad in the late third millennium B.C. In Babylonian myths Anu, the god of heaven, was the highest of the gods, representing majesty and absolute authority. Enlil, second among the gods, was the god of storms, master of the space between heaven and earth, personifying violence, force, and compulsion. The earth or earth-god appeared in several different forms, chief of which was Enki.

In the *Enuma Elish,* we have a detailed cosmogony that describes the origin of the world in a conflict between the forces of chaos and the gods. The primeval chaos out of which all is generated was Nammu in the earliest Sumerian myths, but in the *Enuma Elish* it is divided between Apsu, the sweet or fresh waters, and Tiamat, the salt or sea waters. Out of a union of the fresh and sea waters came the pair of gods Lakhmu and Lakhamu, apparently representing silt or sediment in some sense. Here the geography of lower Mesopotamia shapes the character of the creation myth, reflecting the deposition of silt from the rivers at the point where the fresh and salt waters meet in the Persian Gulf.[11]

Lakhmu and Lakhamu gave birth to Anshar and Kishar, two aspects, upper and lower, of the horizon circle, formed of sediment. Anshar and Kishar gave birth to Anu, god of the sky, and Enki, god of earth. As with the other Mesopotamian gods (figs 2.4, 2.5), Enki was not merely god of earth, he was earth-god, the personification of earth. Conflict arose between the forces of chaos—Apsu, Tiamat, and Mummu—and the younger gods, which led to the slaying of Apsu by Enki (the earth swallowing up the fresh waters). Tiamat was slain by Marduk, a younger god, who was the principal god of Babylon. Marduk, after killing Tiamat, sliced her body in half, and raised the upper half to form the heavens, supported by the wind or storm. The final victory over chaos had been won. Marduk then established the stars and constellations, the sun and moon for the months and days.

Figure 2.4 Babylonian cylinder seal impression ("seal of Adda") dating from about 2200 B.C. The sun god Shamash is seen rising from between the mountains in the bottom center. Ishtar (Venus), goddess of fertility and war is to his left and the water god Ea is on the right. British Museum.

Figure 2.5 Babylonian boundary stone of Nebuchadnezzar I, 12th century B.C., from Kudurru. In the upper register are seen the eight-pointed star of Venus-Ishtar, the crescent of the moon god Sin, and the sun god, Shamash. J. Campbell, *The Mythic Image* (Princeton: Princeton University Press, 1974) and British Museum.

In reciting and dramatizing the *Enuma Elish* at the New Year's Day Festivals (which occurred at the vernal equinox), the Babylonians participated in the victory over chaos, an important psychological event in the life of primitive man.

Egyptian Cosmology

Egypt is, except for the fertile strand of the Nile Valley, a vast and unchanging desert. Not surprisingly, the sun-god became the supreme, the creator god in Egyptian mythology. The day–night death and re-birth of the sun played a central role in Egyptian myth, paralleled by the annual rise and fall of the Nile. The Nile and its flooding were the essence of Egyptian civilization; without it there would have been noth-ing but nomadic tribes. As in other archaic civilizations, the crucial role of water, whether of the sea or flowing streams, in the genesis and sustenance of life was obvious. For the Egyptians life had been created out of the primeval ocean, Nun.

The sun-god Ra took on a number of complementary, rather than conflicting aspects, embodying the differing roles attributed to the sun— a soaring falcon, a scarab beetle rolling the solar disc over the vault of the heavens; as Ra and Ra-Atum at Heliopolis, as Amon-Ra, the Im-perial god of Thebes.[12]

The earth was a flat disc or platter, Geb, surrounded by a mountain-ous rim and floating on the abyssal waters Nun. Out of this primeval sea the land and life appeared, first upon a primeval hillock. Over the

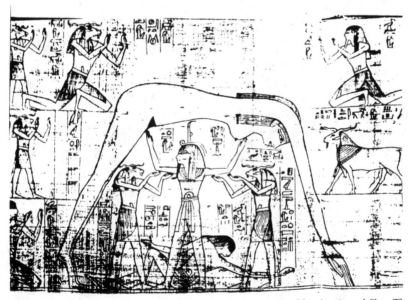

Figure 2.6 The Egyptian sky goddess Nut, her body supported by the air god Shu. The earth god Geb reclines at his feet. From the Book of the Dead, Deir el-Bahri, 10th cen-tury B.C. British Museum.

earth was the dome of the sky, personified as the sky-goddess Nut. The vault of the heavens was often seen as formed by the body of Nut (fig. 2.6), with the stars and planets suspended from her body as ornaments, or Nut was seen as the belly of a cow. In either case the boat of the sun traversed the back of Nut on its daily course through the sky. The heavens were sometimes seen as supported by posts at the four corners of the world, or by the god of the air, Shu.

The Egyptian universe was orderly in its conception, but there was no single dominant or unique cosmogony. Heliopolis, Memphis, and Thebes each had its own cosmogony, its own creator-god who ultimately came to represent a solar deity or created the sun-god. At Heliopolis it was Atum, the primeval god, at once all and nothing, the supercharged vacuum, and Ra-Atum, the sun-god. From Atum came Shu and Tefnut, air and moisture, and from the mingling of air and moisture came Geb and Nut, earth and heaven. Geb and Nut gave birth to the gods Osiris and Isis and Seth and Nephthys. At Thebes, in the Middle Kingdom, the principal god was Amon, who became Amon-Ra, king of the gods. At Memphis, it was Ptah, who created all from his own breath.

Conclusion

Two elements will be emphasized as we follow the development of cosmology. First there is the organization of the world, which was embodied in heroic myth for the Middle Eastern peoples. The mythopoeic (mythmaking) view of the world was to be altered beyond recovery by the Greeks, who invented the rational methods of philosophy. The second element is number. The Egyptian skill with applied geometry is evident in the architecture of their monuments, and both Egyptians and Mesopotamians maintained accurate calendrical records over long periods. The Babylonians have left us a long record of astronomical observation, motivated in the early period by religious considerations, later by astrology and a growing interest in mathematical astronomy. Yet the mathematics did not truly influence their world view, and it remained for the Greek philosophers to sense that the world cannot be understood without recourse to the quantitative. For this they needed the meticulous records of the Babylonians, especially the observations of the positions of the planets, which became elements in the first physical cosmologies.

CHAPTER 3

First Astronomy

The Riddle of Time

We have only the sketchiest information about the dawning awareness of celestial order among early civilizations. For primitive man, as
for us, the dominant regularity was the day-night cycle; as the sun defined the day, so the moon ruled the night. Its importance in illuminat-
ing the frightening darkness gave the moon a crucial place in man's
superstition. Among certain peoples it was worshiped as a living being,
and its reappearance each month was celebrated with offering and cer-
emony. There is evidence that upper Paleolithic man was concerned
with the cycle of the moon and was recording the number of days
between one new moon and the next.[1] Some of these tantalizing artifacts, which represent not only our earliest evidence of human interest
in the skies but also the first evidence of tallying or counting of any
sort, are shown in figure 3.1. Along with the renowned cave paintings
of the same period, which show a highly developed aesthetic sense,
these discoveries give us just a glimpse of the intellect of early man.
Thus perhaps 30,000 years ago primitive man was not only aware of
the lunar cycle but also was apparently already using it to keep track of
the days, noting times of invisibility and indicating the changing phases
of the moon. Furthermore, he was acutely aware of the alternation of
the seasons between the dead winter and abundant summer or between
dry and rainy seasons; the life cycles of animals on which early man
depended were seasonally regulated, as were the wild seeds or grain he
gathered. When early nomadic peoples or fishermen began to travel
widely, a new importance of the astronomical regularities emerged; they
were relied upon for orientation and navigation. The stars were used to
tell the time of night, their rising and setting points served as a com-
pass. These nomadic groups first regulated their primitive "calendars"
entirely by the 29½ day synodic period of the moon (the interval be-

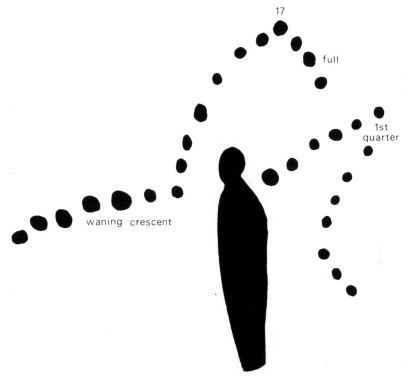

17

full

1st
quarter

waning   crescent

Figure 3.1A   The moon god (?) and record of the days of the lunar month from the
paleolithic of Abris de las Vinas, Spain. A. Marschak, *Science* (1964) 146:743. Copyright
© 1964 by the American Association for the Advancement of Science.

tween one full moon and the next); "The Moon has been created for
the counting of the days," says the Hebrew Midrash. Eventually it became apparent that the seasons repeated after approximately twelve lunar cycles; the month of rains, of sowing, or of harvesting returned.
Agricultural people were strongly tied to the solar year, or as we call it
the tropical year, especially in high latitudes, where seasons were distinct. This protoastronomy emphasized one aspect of what was to become science—the computational aspect, the study of regularities. Time
reckoning and navigation were to provide much of the impetus toward
understanding of astronomical cycles.

In trying to understand the accomplishments of prehistoric and early
historic protoscientists, we are faced with the difficulty of appreciating
their ignorance of the world. As Toulmin and Goodfield have written,
"they were confronted not by unanswered questions, but by problems
as yet unformulated, by objects and happenings which had not yet been

set in order, far less understood."[2] The very classification of phenomena, the understanding that rain, lightning, locusts, comets, and other
things that appear in the sky are essentially dissimilar, was achieved
only after millennia of misunderstanding. As late as 700 B.C. Mesopotamian sky watchers carefully recorded eclipses, clouds, and lunar halos
as happenings of the same kind. All were omens, to be interpreted for
their religious or civil significance. It is in this sense that the meticulous
observations of the Babylonians are described as "pretheoretic." The
regular was as mysterious in its origin as were the irregular and non-
recurring.

Much of the motivation for recordkeeping over long periods was the
need for an adequate calendar. A basic difficulty for any calendar giver
has already been remarked: The solar year of 365¼ days is incommen-
surable with the lunar month. And neither period is an integral number
of days. As a consequence the ages have seen many calendars, none of
them fully satisfactory, even to the present day.

The Egyptians maintained two quite independent calendars. Their lunar or astronomical calendar had a year of 365¼ days beginning with
the Sothis, the heliacal rising[3] of the star Sirius, the brightest star in the
heavens, which occurred near midsummer and coincided with the rising of the Nile. The Egyptian civil calendar, though having obvious
solar and lunar bases,[4] was essentially nonastronomical; it consisted of
12 months of 30 days each, plus 5 days at the end of the year. The
historian Otto Neugebauer has characterized this calendar as "the only
intelligent calendar which ever existed in human history."[5] Certainly
the contrast is dramatic between the Egyptian civil calendar and other
calendars of antiquity, typically based on the lunar month with inter-
calations (added days or months) that were often made quite haphaz-
ardly. The first of Thoth, "New Year's Day" on the Egyptian civil
calendar, evidently coincided with the Sothis at some early date when
the two calendars were regularized. Thereafter the two calendars got
out of phase, since the civil calendar gained one day in 4 years and one
full year approximately every 1,460 years (365 × 4). This has been useful in fixing early Egyptian chronology, since the two calendars apparently coincided at about 2773–2770 B.C. The Egyptians divided the zodiac into 36 "decans" of 10° each, three to a constellation of the zodiac.
The year had three seasons of four months each, a division based on
the cycle inundation, sowing, and harvesting. From the division of the
8 hours of darkness at the summer solstice into 12 decans came the 12-
"hour" night, and eventually the 24-hour day.[6]

The Babylonians never departed from a purely lunar calendar; the
month began at the first visibility of the crescent moon. The use of a
strictly lunar month led to alternating months of 29 and 30 days. A

year of 12 months had but 354 days, so that every three years the calendar was off by more than one month. For this reason a cycle of intercalation was adopted in which one month was added every three years, leaving a residual error of only three days. A better approximation was an eight-year period of 99 lunar months, leaving an error of less than two days. This cycle was used by the Assyrians and Chaldeans as late as 480 B.C. A century later the 19-year, or *Metonic* cycle, based on the equation [19 solar years = 235 lunar months], was in use, with seven 13-month years in 19. The 19-year period became the basis for the Jewish calendar. For religious and political reasons Mohammed restored the 354-day year of 12 lunar months. This return to a primitive calendar reduced the ties of Islam to the Jews and also took away the prerogative of the sheiks to determine whether there would be an intercalary month in a given year.

Pretheoretic Astronomy

Although astronomy as a systematic observational endeavor originated in Mesopotamia, it is important to distinguish between the largely primitive efforts of early Babylonian astronomy and the serious observations and calculations of the Neo-Babylonians and Persians, which were contemporaneous with the highest achievements of Greek science. Studies by Neugebauer[7] and others make clear the primitive and qualitative character of early Babylonian astronomy and show that even Assyrian and later astronomical efforts were at least as concerned with mathematical theory as with accurate observation. In the *Gilgamesh* and the *Enuma Elish* epics, references to the sun, moon, and stars are not unlike those to be found in the Hebrew Genesis creation myth. On the other hand, the observations of Venus found in the great library of Ashurbanipal (669–627 B.C.) permit astronomical dating of the period immediately following 2000 B.C.

In Assyrian times truly systematic observations began to be made, so that Ptolemy in A.D. 150 could claim knowledge of eclipses continuously since 747 B.C. From about 700 B.C. continuous observations of the planets, especially their heliacal risings and settings, were made. The primary motivation for these observations seems to have been astrology. Here is an example of an Assyrian astrological text:

Venus disappears in the West. When Venus grows dim and disappears in Abu there will be slaughter in Elam. When Venus appears in Abu from the first to the thirtieth day, there will be rain, and the crops of the land will prosper. In the middle of the month Venus appears in Leo in the East.[8]

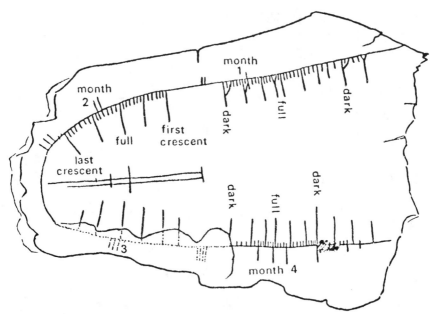

Figure 3.1B Lunar "tally-marks" on a fragment of late upper paleolithic mammoth tusk from Gontzi, the Ukraine. From G. Hawkins, *Beyond Stonehenge*, after A. Thom. By permission of Harper & Row, Publishers, Inc.

After the fall of Ninevah in 612 B.C., Babylon became the capital of the New Babylonian Empire and the center of astronomical activities. In cuneiform astronomical texts from the period, many of the constellations are recognizable, and far more important, the periods of the planets are employed in calculation. By the fourth century B.C. the zodiac was in use and the "Saros" cycle of 223 lunations between essentially identical series of eclipses had been discovered. By 300 B.C., after the death of Alexander, Babylonian mathematical astronomy reached its zenith.

As far as is known, the only astronomical instrument possessed by the Mesopotamians was the gnomon, a sort of vertical sundial consisting merely of an erect stick in an open, flat area (fig. 3.2). With a gnomon one can determine the solstices from the extreme positions of the sun's shadow, the equinoxes when the sunrise and sunset positions are opposite each other. Indeed, with this device one can establish all the essentials of a solar calendar. Furthermore, north-south is given by the average direction of the sun's shadow at noon, when the shadow is shortest. Although angular positions of the less luminous heavenly bodies can be measured with instruments almost as simple as the gnomon (it is only necessary to sight over the stick rather than depend on the

Figure 3.2 A gnomon in use by Bornean natives. Note the use of small sticks to measure the length of the shadow. C. Hose and W. McDougall, *The Pagan Tribes of Borneo* (New York: Macmillan, 1912).

shadow cast by it), as far as we know the Mesopotamians developed no such instruments. They therefore recorded those positions that can be obtained without instruments: risings and settings, eclipses, and other conjunctions. Observations near the horizon are more often impeded by poor visibility than those at higher altitudes. Consequently, accurate values for the intervals between successive occurrences of the same celestial event come only from long observation over many repetitions of the event.

The need to understand the regularities hidden in the tables of the Babylonian astronomers gave impetus to the development of mathematics. The mathematical treatment, like the astronomy itself, was pretheoretic. Phenomena in the sky were examined for patterns, but it was only for astronomical events that precise correlations could be found. The equinoxes always succeed each other in a regular way. But even the simplest motions of celestial objects are not quite uniform. The sun, for example, moves through the stars somewhat faster in winter than in summer. The Babylonian mathematicians dealt with this problem by computing the differences of actual cycles from ideally repetitive cycles. The sexagesimal (base 60) number system of the Babylonians was a "place value" system that made it possible to deal accurately with the small numbers that resulted when successive differences were taken. By contrast the Egyptian mathematics was unsystematic and cumbersome. Perhaps the inadequacy of the Egyptians' number system was in itself sufficient to account for their failure to match the achievements of the Babylonians. Otto Neugebauer's conclusion is that "mathematics and astronomy played a uniformly insignificant role in all periods of Egyptian history."[9]

Many of the most impressive advances of Babylonian protoscience occurred within a relatively short period during the first millennium B.C. Intense interaction between mathematics and astronomy, to the benefit of both fields, is encountered again in the scientific achievements of the seventeenth century, which capped the "Copernican Revolution." It is plausible to speculate that in the Babylonian "prescientific revolution" there must have been at work men whose intellects equaled those of Aristotle, Newton, and Einstein. But these earlier heroes must remain nameless.

Archaeoastronomy

Up to this point we have focused our attention on early Babylonian and Egyptian astronomy for several reasons, not the least of which is the crucial importance of these cultures in the development of Western civilization. Certainly the earliest astronomical texts are found on Babylonian cuneiform tablets dating from almost 2000 B.C., and interest in calendrical questions in both these civilizations led to consideration of basically astronomical problems. Many of our constellations were first named by the early Mesopotamians, and Egyptian representations of the heavens (fig. 3.3) not only are decipherable but also clearly show the role of the stars in Egyptian religious thought. During the first millennium B.C., Babylonian astronomy began to take on a complex mathematical character and much effort was devoted to the problems of astronomical prediction.

Thus, not so very long ago, a discussion of pretheoretic astronomy ("protoastronomy") would have concluded here, perhaps with a brief comment on the elaborate calendar of the Maya of the new world. In the last two decades, however, astonishing evidence of man's early

Figure 3.3 Constellations of the Egyptian northern sky from the tomb of Seti I at Thebes. The composite hippopotamus/crocodile is thought to represent Draco, or a part thereof. The "hitching post," on which the composite animal rests a hand, is possibly a manifestation of the pole. The big dipper (foreleg, plow) is pictured as a bull. R. A. Parker, "Ancient Egyptian Astronomy," in F. R. Hodson, *The Place of Astronomy in the Ancient World* (London: Oxford University Press, 1974).

preoccupation with the heavens has come to light, and in such profusion that an entirely new perspective, "archaeoastronomy," is required. Much of the interest in this field dates from Gerald Hawkins' writings about Stonehenge, which captured the popular imagination. Hawkins' studies and those of his contemporaries awakened professional anthropologists and archaeologists to the astronomical content of ancient sites, which range from Egypt to Cambodia, and from Scotland to Peru. Some of the groundwork was done during the last century, but much of it was considered irrelevant or controversial. Only recently have the combined techniques of archaeology and astronomy been systematically applied, especially in Britain and in the New World. From these efforts has emerged a picture of an enormous preoccupation with the heavens in many early cultures, manifested in differing ways in different times and places, but with many elements in common, among both literate and preliterate peoples. It is this story that we shall endeavor to tell. We begin with Stonehenge, not because it is unique, but because of its hold on the imagination and the thoroughness with which it has been studied. It is the most dramatic, the most evocative of myth and speculation, and the most familiar of all the ancient astronomical sites,[10] and much is to be learned from a detailed examination of it that will carry over into our discussion of the diverse sites found over much of the world.

Sometime in the third millennium B.C., before Hammurabi ruled
Babylon and contemporaneous with the Middle Kingdom of Egypt,
early Britons began building the dramatic stone monument we now
know as Stonehenge (fig. 3.4). It has long been known that Stonehenge
was oriented in the direction of sunrise at the summer solstice. As long
ago as 1740 William Stukeley speculated about the astronomical importance of Stonehenge, and in 1903 Lockyear drew attention to its astronomical alignments. But only in very recent years, through the efforts
of C. A. Newham, Gerald Hawkins, and Alexander Thom, have we
learned that not only Stonehenge but also hundreds of other megalithic
monuments in Great Britain and on the continent exhibit a great variety
of astronomical alignments, displaying a sophistication far beyond that
expected of the early inhabitants of these areas.

Modern archaeology has been able to distinguish three stages in the
construction of Stonehenge, spanning the period 2750–2075 B.C.[11] During the first phase, Stonehenge I, the great circumscribing ditch and
banks, 380 feet in diameter and open on the northeast (fig. 3.5), were

Figure 3.4 An aerial view of the ancient megalithic monument of Stonehenge showing
the outer sarsen circle and the inner horseshoe of massive trilithons. G. Hawkins, *Vistas
in Astronomy* (1968) 10:45.

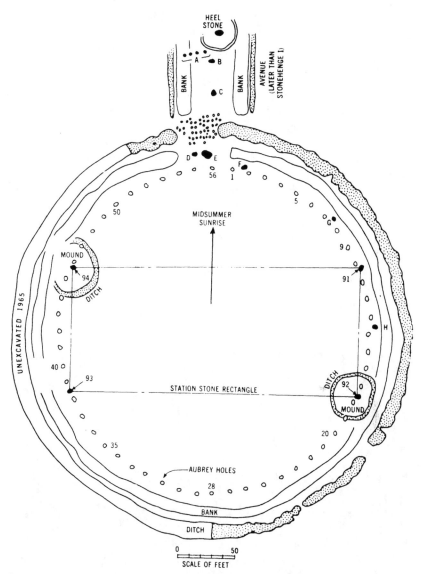

Figure 3.5 Plan of early Stonehenge (Stonehenge I, with features of Stonehenge II), showing the heel stone at the top, the four station stones (91–94), and the Aubrey circle of 56 holes. G. Hawkins, *Vistas in Astronomy* (1968) 10:45. By permission of Pergamon Press Ltd.

dug, as were the 56 so-called Aubrey Holes, of which we shall hear more later. Several stones were placed in position during Stonehenge I, including the important "heel stone" (fig. 3.6). It may also be that the station stones, numbered 91–94 in figure 3.5 were erected at this time. The second wave of building began about 2150 B.C., resulting in Stonehenge II; the builders were the early Bronze Age people known as the "Beaker people." More than 80 megaliths called bluestones and weighing up to 5 tons apiece were erected in two concentric circles near the center of the enclosure, and the northeast opening was made into a well-defined avenue. Shortly thereafter, beginning about 2075 B.C., the construction of Stonehenge was largely completed by the Wessex people, who replaced the double circle of bluestones with a circle of 30 "sarsen stones" with lintel caps, and the giant horseshoe of trilithons, which dominates the present striking ruin. The trilithons were made from stones 25 to 30 feet long and weighing as much as 50 tons. Finally some of the bluestones were reerected inside the horseshoe, and the 59 "Y" and "Z" holes were dug.

Let us look at Stonehenge in detail, following Hawkins, and consider the evidence that it was a solar and lunar observatory. The declination of the sun, that is, the angle north (+) or south (−) of the plane of the earth's equator, varies from +23½° at the summer solstice to −23½°

Figure 3.6 The sun rising over the heel stone on midsummer's day, as seen from the center of Stonehenge. Courtesy of CBS Television and Souvenir Press.

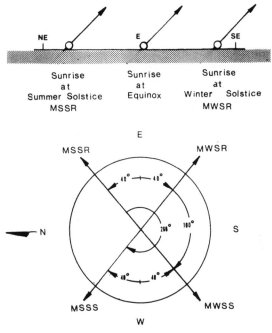

Figure 3.7 Above: Sunrise at the equinoxes and solstices, showing the effect of the sun's declination on the rising point. Below: Plan view showing sunrise and sunset at the solstitial extremes.

at the winter solstice (See figure 2.2B). As a result the sunrise point at a northern latitude ranges from well to the north of east in summer to south of east in winter. Similarly, the sunset position varies from north of west to south of west (fig. 3.7). At the solstices the sun rises and sets to the north or south by an amount that depends on one's latitude. Stonehenge is at latitude 51°N, so that at the summer solstice the sun rises 40° north of east and sets 40° N of west, whereas at midwinter the rising and setting points are 40° to the south of the east and west points, respectively.

In 2000 B.C. the declination range of the sun was ±24,[12] so that the amplitude of the sun at the summer solstice was almost 41°, nearly northeast at sunrise or northwest at sunset. The avenue at Stonehenge is oriented only approximately along the direction of sunrise at the summer solstice (midsummer sunrise, MSSR), but the line from the center to the heel stone is almost exactly the MSSR direction for declination +24°. Figure 3.8A shows some of the other significant solar alignments.

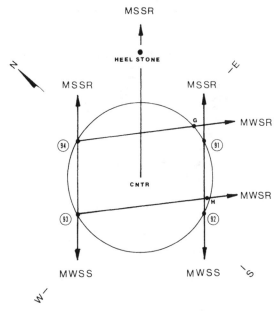

Figure 3.8A Solar alignments at Stonehenge. MS and MW denote midsummer and midwinter, respectively, and SR and SS denote sunrise and sunset. At midsummer (summer
solstice), the sun's declination is +23½° (+24), while at midwinter (winter solstice) the
declination is −23½° (−24).

Because they were unsuspected before Hawkins' work, the lunar
alignments are of greater interest. The inclination by about 5° of the
moon's orbit relative to that of the earth means that the extreme declinations of the moon can range as much as 5° greater or less than those
of the sun. Thus at Stonehenge in 2000 B.C. the lunar excursion in declination ranged from ±19° at the smallest, to a maximum of ±29°. The
corresponding excursions on rising or setting are ±30° to ±51°. The
moon might rise as much as 51° north or south of east, and so on.
Although among primitive peoples new moon ordinarily exceeded full
moon in importance, moonrise and moonset cannot be observed for
the new moon. So it is reasonable to concentrate on the rising and
setting of the full moon at Stonehenge and in particular on the solstitial
rising and setting, that is, on midwinter moonrise (MWMR), midwinter moonset (MWMS), midsummer moonrise (MSMR), and midsummer moonset (MSMS). Figure 3.8B shows a number of the lunar alignments at Stonehenge. Hawkins has also identified solar and lunar
alignments involving the trilithon and sarsen archways. Thus a number
of the features of the site are clearly identifiable as sighting points for

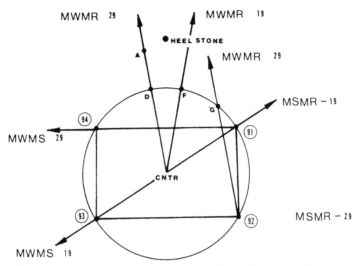

Figure 3.8B Lunar alignments at Stonehenge, according to Hawkins. Midwinter moonset is denoted by MWMS, and so on. The extreme declinations of the moon range from +19° to +29° in an 18.61-year cycle.

observing the rising and setting of the sun and moon. Not all of the alignments are independent, of course, since, for example, the directions MWSS and MSSR are opposite each other. Moreover, since the sunrises and sunsets range from 40° S to 40° N, and since the largest of the lunar extremes (±29°) correspond to N–S swings of ±51°, some of the solar alignments are almost perpendicular to some of the lunar alignments. This gives rise to fortuitous sight lines in a structure with right angles built into it. Thus MWMR (+29) and MWSR (−24) or MSSS (+24) are in directions perpendicular to each other, so that if the station stones are set in a rectangle, with one pair of sides along the MSSR–MWSS direction, then the other will automatically define the directions MWMS (+29) and MSMR (−29). A rectangle is formed in this way only at the latitude of Stonehenge, plus or minus a degree or so; elsewhere a parallelogram would result. It may be that the location of Stonehenge was chosen for that reason. On the other hand, it has been suggested that the station stones were simply set in a rectangle and that the lunar alignments were accidental. The existence of other lunar alignments such as center to D and A (MWMR [+29]) and center to F (MWMR [+19]), etc., seriously weaken that argument. Most important, the alignment 91-93, along the MSMR (−19) directions, can hardly have been fortuitous and in fact determined the shape of the rectangle.

British anthropologists were originally reluctant to accept much of

Hawkins' discussion and considerable criticism was leveled at his statistical analysis of the calculated correlations and at the manner in which he selected features among which to make his computer-aided search for alignments.[13] Although some of the criticisms were valid, they do not destroy his general thesis. Recently Newham[14] has noted that a chain of filled holes centered about the MSSR direction seem to represent the process of following the regressing moon as it swung from one extreme to the other and back again. Thus it may be that these post holes provide us with the opportunity to "watch" the builders of Stonehenge at work, as they follow the swing of the moon from one extreme to another, over a century or so.

More controversial still is Hawkins' further suggestion that Stonehenge not only served as a device to mark the solstices and other important solar and lunar phenomena and, of course, as a stage for associated celebration or worship, but also functioned as an eclipse predictor! Nothing that was previously known of the early Britons suggested such a preoccupation with or understanding of solar and lunar events, and we know that Babylonian astronomers of a much later period were unable to predict eclipses. The suggestion is thus a radical one and has met with considerable skepticism from anthropologists. In support of Hawkins' thesis, some of the work of Thom indicates an extraordinary understanding of lunar phenomena by these British megalithic astronomers—a knowledge of exactly the sort that would make eclipse prediction possible. Thus Hawkins' proposals are at the very least deserving of close scrutiny, as examples of how Stonehenge might have been used to predict solar and lunar eclipses.

In an article entitled, "Stonehenge, a Neolithic Computer,"[15] Hawkins claimed that the 56 Aubrey holes were used as a computer to predict "danger periods," when eclipses might occur. The period of regression of the nodes (the points at which eclipses can occur) of the moon's orbit is 18.61 years. That is, every 18.61 years the entire orbit turns around completely in space. This means that the points in the sky where the paths of the sun and moon cross move westward in the sky at a rate that takes them around the entire ecliptic in 18.61 years. The "eclipse year," the time between successive passages of the sun past the same node, is 346.62 days (about 19 days less than the tropical year). A solar or lunar eclipse can occur near each of the two nodes, that is, about every 173 ± 15 days. (In fact, a solar eclipse *must* occur somewhere on the earth.) Now 223 lunar (synodic) months equal almost exactly 19 eclipse years and very nearly 18 tropical years (18 years 11 days). *Similar* eclipses then recur with a period of 18 years and 11 days, known as the "saros." We know that the Babylonians had knowledge of this 18-year cycle somewhat after 600 B.C. Hawkins supposes that the early Britons discovered an eclipse cycle of 56 years, which is a consequence of the fact that 59 eclipse years almost precisely equal 56 tropical years (or essentially that three cycles of regression of the nodes equals 55.8 years) and eclipses then recur at the same time of the year at a given location. Then three cycles of regression could be represented, integrally, by intervals of 19, 19, and 18 years, totaling 56, with an average of 18.67. Indeed, a particular phenomenon, say full midwinter moonrise occurring at an extreme, will occur at intervals of either 18 or 19 years, 19 occurring slightly less than two-thirds of the time. An eclipse of the sun or moon always takes place (within 15 days of the solstice) when MWMR occurs over the heel stone, since this indicates the moon is at a node. Fewer than one-half of the eclipses predicted by this method would be visible from Stonehenge itself, but the use of the 56-year cycle would predict when an eclipse could occur.

In any event, a possible explanation for the puzzling number of 56 Aubrey holes has been advanced. It is at least plausible, though supported by no direct evidence, and it does succeed where others have failed, in explaining one of the principal anomalies of Stonehenge. The fact that the Y and Z holes number 30 and 29 respectively and were almost certainly used to count the days of the lunar month (29½ days) in a manner quite close to that suggested for the Aubrey holes and eclipse prediction adds credibility to his thesis. Hawkins has been joined by others, notably Fred Hoyle, in proposing other eclipse prediction schemes using the Aubrey holes.[16] Hoyle, indeed, has argued that since eclipse prediction is such a complex process, that we can use the Aubrey holes to predict eclipses is prima facie evidence of their original purpose. The Aubrey holes date from the Stonehenge I phase, which requires that one imagine that active solar and lunar observations, perhaps on the same site, had already carried out for many decades.

As we have already noted, Hawkins' ideas (and those of Newham) were not received with enthusiasm by British anthropologists. They seemed to bespeak a sophistication on the part of these preliterate stone age and early bronze age Britons hitherto unsuspected. But we shall see that the preoccupation with solar and lunar phenomena at Stonehenge was by no means isolated, that scores of astronomical sites are scattered over the British Isles, and that other cultures worldwide have used "horizon astronomy" in a similar fashion. In any event, it now seems clear that Stonehenge was used as an astronomical observatory and that it was employed to fix the dates of the solstices, probably for ceremonial purposes. A great interest in lunar phenomena has also been demonstrated by the existence of important lunar sight lines, and it does not take a great leap of the imagination to see Stonehenge as an eclipse computer.

Other Megalithic and Ancient Astronomical Sites

While Hawkins' writings were creating a controversy over the nature of Stonehenge, other scholars, notably the English civil engineer Alexander Thom, were publishing studies of sites of similar character in Britain that clearly exhibit alignments of an astronomical origin and evidently were calendrical in nature. Thom has demonstrated convincingly that a large number of British megalithic sites of great variety in shape and complexity were employed in solar, lunar, and stellar observations. Hundreds of megalithic sites were constructed from the late third millennium B.C. onward, perhaps over a period of a thousand years. Those that may have astronomical content consist of standing stones (menhirs) and stone circles. Some clearly had as part of their purpose the making of solar or lunar observations, while others were primarily ritualistic or symbolic. Many, if not all, were associated with burials. Furthermore, Thom's work suggests that eclipse prediction may have been one of the uses of some of these constructions. Most of the sites Thom identified involve the use of natural foresights, usually features on the distant horizon. More strongly ceremonial sites like Stonehenge, using standing stones to determine solar and lunar alignments, necesssarily suffered in accuracy by comparison.

The reasons for the great preoccupation with lunar phenomena on the part of the early Britons remain obscure; an understanding of the tides was important, of course, as Thom points out in the introduction to *Megalithic Lunar Observatories,* and eclipses were no doubt the bearer of the most powerful portents. Still, the full story will never be known.

One has only to peruse Thom's works or the rapidly growing literature on paleoastronomy in pre-Columbian Middle America to see how pervasive this pretheoretic interest in positional astronomy was. Astral-lore is, of course, an element in the mythology of almost all cultures, but our interest here is in the attention given to rather specific and technical problems associated with regularities of the sun, moon, and stars. Typically, as with Stonehenge, the astronomical alignments are toward rising and setting points of bright celestial objects.

Among other examples of astronomical sites from the British Isles are the stones at the controversial site of Callanish in Scotland (fig. 3.9), whose alignments may have marked the rising of bright stars, or perhaps the sun and moon; and the sites of Kintraw, Ballochroy, and

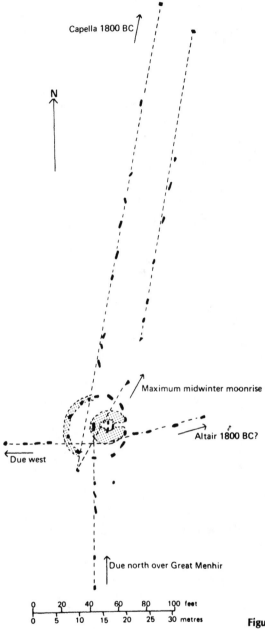

Capella 1800 BC

N

Maximum midwinter moonrise

Altair 1800 BC?

Due west

Due north over Great Menhir

0   20   40   60   80   100 feet
0   5   10   15   20   25   30 metres

Figure 3.9 Callanish. This dramatic megalithic site in the Outer Hebrides has rows of standing stones which suggest astronomical alignments. B. Sommerville, *Journal Royal Anth. Inst.* (1912) 42:23.

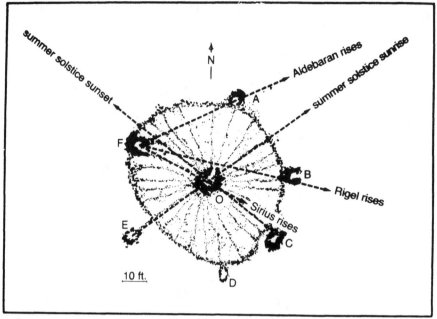

Figure 3.10 Apparent astronomical alignments at the Big Horn Medicine Wheel in Wyoming. J. A. Eddy, *Science* (1974) 184:1035.

Woodhenge.[17] In the vicinity of Carnac in Brittany are found incredible arrays of standing stones, exemplifying an extraordinary preoccupation with stone circles and lines of menhirs; thousands of stones are involved.

In America, where every American Indian pueblo even now is said to have an office of sun-watcher, solar shrines have been identified at many sites. Among them are Chaco Canyon in New Mexico, and the Zuni pueblo, Arizona, where a sun-watching tower is known. At Canyon de Chelley many sacred planetarium sites are known, and elaborate star patterns have apparently been identified in petroglyphs from the Great Basin in western North America; the eleventh-century Casa Grande ruin shows many astronomical alignments, as does the Big Horn Medicine Wheel (fig. 3.10) in Wyoming. In Middle America there are suggestions of Pleiades[18] orientations at Teotihuacan, while a building at Monte Alban seems to be oriented toward the rising point of the bright star Capella, whose heliacal rising in 275 B.C. was at about the time of the first zenith passage of the sun there. At Caballito Blanco there appear to be MSSR and Sirius alignments. The "Caracol" at the Mayan ruin of Chichén Itzá in Yucatan is the most famous "observatory" in meso-America, with solar, lunar, and Venus alignments for

(a)

Dresden Codex

(b)

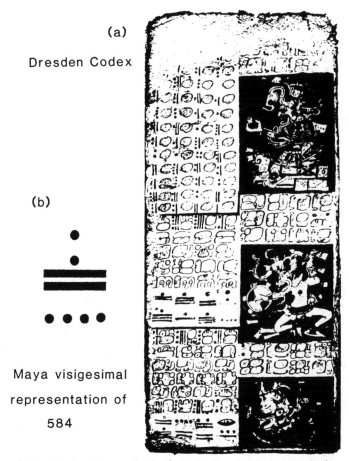

Maya visigesimal

representation of

584

Figure 3.11 (a) Maya Venus calendar from the Dresden Codex. At the lower left are Maya visigesimal representations of 236, 90, 250, and 8 days which are the approximate values of the periods during Venus' 584 synodic period when it is (1) the morning star, (2) when it disappears "behind" the sun during superior conjunction, (3) the 250 days of appearance as evening star, and (4) 8 days of invisibility. For further details of this page of the Dresden Codex, and the next four pages of the Venus tables, see J. E. S. Thompson, "A Commentary on the Dresden Codex," *Mem. Am. Phil. Soc.* (1972):93 or Thompson, *Maya Hieroglyphic Writing* (Norman: University of Oklahoma Press, 1960). (b) Maya visigesimal (base 20) representation of 584. With a dot representing 1 and a bar denoting 5, the diagram reads, from bottom to top, $4 \times 1 + (5 + 5 + 1) \times 20 + 1 \times 360 = 584$. The top line represents a multiple of 360 (in this case, 1) rather than 20^2.

A.D. 1000; there may also be Venus orientations at Uxmal. At Uaxactun there is a solar "observatory," while at Copan there is a sightline denoting the helical rising of Capella, which coincided with the date of the first zenith passage of the sun. Some of these alignments involve overall orientations of buildings, or walls, others use cross-jamb or parallel-jamb orientations of windows or other wall openings. Among effigy-mounds in North America are shapes that may represent stars or star patterns, perhaps in the sense of a template matching. (It has been speculated that the famous "sand-drawings" on the Nazca plain in Peru might have a similar purpose, perhaps mirroring the Milky Way). There are remarkable Venus tables in the Mayan Dresden Codex (fig. 3.11) and eclipse tables as well.[19] These also date from about 1200. Venus glyphs adorn the Temple of Venus at Copan. The Mayan fascination with regularities and cyles is dramatically expressed in their calendar. The odd 260-day period (tonalamatl) has no clear astronomical basis, being the product of 20 (day names) and 13 (day numbers). There was an approximate year of 360 days (tun), with 18 divisions of 20 days each. The "vague year" of 365 days consisted of one tun and a remainder of five days which were considered unlucky. Cycles of 400, 400^2, and 400^3 (64 million) tuns are known.

Finally, several Egyptian monuments and temples show orientations toward the midsummer and midwinter sun, including the temple of Amon-Ra at Karnak, and perhaps toward the first crescent moon following the summer solstice.

This litany of the astronomical and calendrical information stored in so many ancient monuments around the world should persuade even the skeptic that primitive man's interest in the heavens was not confined to star-lore or myths of creation or expressed merely in a vague qualitative wonderment at the mysteries of the skies. For although we may never have a deep understanding of why a monument like Stonehenge was built, we now know that all over the world, careful, continuous observations of certain astronomical phenomena were being made and the information was "stored" in monuments and buildings.

CHAPTER 4

The Fountainhead

Origins

During the Homeric period (900–700 B.C.), semipastoral barbarians invaded the eastern Mediterranean, destroying the highly developed civilizations of Knossus, Mycenae, Pylas, and other cities. The light of civilization flickered only dimly during this period, which was comparable in some ways with the "Dark Ages" of medieval Europe. Gradually the early Hellenic civilization emerged, and the cities of Ionia in Asia Minor became a center of commerce in touch with Egypt, Babylonia, and other ancient civilizations. It is perhaps not surprising that it was here, in the Ionian city of Miletus late in the seventh century B.C., that Greek philosophy had its origins.[1] Unfortunately, blind fate has largely determined which of the early Greek writings have been preserved. The works of Plato and Aristotle have come down to us essentially complete, while many other philosophers are known primarily from second-hand reports. The pre-Socratic philosophers of Ionia and Southern Italy are known only by hearsay and fragments, and the hearsay is often that of hostile critics. Yet enough of the picture of Greek philosophy of the fifth and sixth centuries B.C. has been passed on that the various ideas and influences can be sorted out.

When we recall that our goal is to trace the development of ideas of the universe, we might at first find little to remark in the cosmological models of the Ionians. Their various descriptions of the world are in most ways not very different from and are by no means more accurate than the Babylonian or Egyptian cosmologies. The Ionians, like their Oriental neighbors, proposed a geocentric world. They placed the inhabited earth at the center of things, with a substantial (whether solid, watery or fiery) celestial region surrounding all. The importance of the Ionians is rather in their method: they were the first to treat nature as an object for speculation. The Middle Eastern cosmologies, embodied in life-pervading myth, required no justification and to question the myths would have been irreligious. In large measure the value of the Ionian philosophy lies not in any answers supplied about the organization of the world, but in the questions asked, the problems posed. Fertilized by the freedom of thought of the seafaring merchants of Ionia, the germ of inquiry that was in their first questions grew quickly, so that philosophy was transformed many times over during the ensuing three centuries. We shall not discuss comprehensively even the main developments in Greek philosophy but will be content to recognize the appearance of those persistent ideas that have influenced science even to the present day.

The first Ionian philosopher of whom anything is known was Thales (632–546 B.C.) of Miletus. He apparently wrote nothing, but some of his ideas were discussed by Aristotle and others. A statesman and sage who traveled widely, Thales' views on nature were preserved only incidentally. If he had not been associated with such men as Anaximander, little perhaps could have been made of his one statement on cosmology (as transmitted by Aristotle), that all matter "is water." In context one can interpret this as referring to a first principle or material cause, although rational discussion of the world-as-object came only with the Ionian successors to Thales.

More is known of Anaximander, a somewhat younger Milesian and a pivotal figure in the development of Greek thought. In his writings he postulated a featureless matrix, called "the Unlimited" or "the Infinite," from which the physical world derives its existence. The visible world evolves from the Unlimited by a principle reminiscent of the mythical creation acts: the separation of opposites. Scholars differ about the precise way that Anaximander's cosmos formed. If a pair of opposites—hot and cold, say—are already present in the featureless Infinite, how can it be said to be featureless? And if thoroughly mixed, in what sense are opposites present? Anaximander gave no explanation. His world system was rooted not in mechanism, but in law. All natural processes, he wrote, are governed by an overriding principle of cosmic justice, or Necessity. Anaximander, in thus denying man's preferred status in nature, in asserting that things happen because they must, opened the door to scientific rationalism. Later philosophers tended to take the more anthropocentric position that Purpose, not blind Necessity, undergrids reality.

Anaximander visualized the earth as a squat column that had been left at the center of the cosmic vortex when fire moved to its natural position, uppermost below the whirling heavens. For Anaximander the earth, being already at the center, had no need of support. The sun, moon, and planets were windows in successive rings of fire that sur-

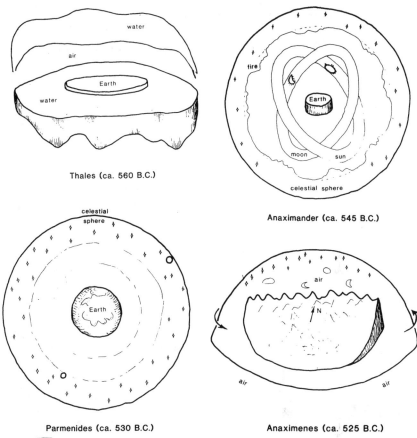

Thales (ca. 560 B.C.)

Anaximander (ca. 545 B.C.)

Parmenides (ca. 530 B.C.)

Anaximenes (ca. 525 B.C.)

Figure 4.1 Four of the earliest Greek cosmologies visualized. (a) Thales; (b) Anaximander; (c) Parmenides; (d) Anaximenes.

rounded the earth "like the bark on a tree."[2] Anaximander's description of the motions of these fiery rings (figure 4.1), and his explanation of eclipses, are vague and unsatisfactory. Nevertheless, the picture of the universe as nested shells foreshadows what Thomas Kuhn calls the "two-sphere universe" (that is, concentric earthly and heavenly spheres). Here is the kind of geocentric universe that, largely because Aristotle adopted it, survived until the seventeenth century.

The Pythagoreans

It was Pythagoras, a compelling but shadowy figure, who saw *number* as the organizing principle of the universe. A native of Samos in Ionia, he settled in Croton, one of many Greek colonies in southern

Italy, in the sixth century B.C. There he founded the Pythagorean School, whose influence was strong for two centuries. Unworldly, disciplined, religious, the Pythagoreans kept their wisdom from the uninitiated. Pythagoras himself wrote down none of his ideas. This fact, and his later followers' habit of attributing everything to the master, compound the mystery surrounding Pythagoras. Certainly he must have been a powerful presence; of the hundreds who accepted his call to a life of contemplation, many are said to have followed without even saying good-bye to their families. The cult they formed, better known than Pythagoras himself, had many features in common with Eastern religions. Transmigration of souls was a central belief. In order to escape from the cycle of reincarnation the Pythagoreans sought purification through music and mathematics and in a rigid personal code that included abstention from eating flesh.

Our particular interest is in the Pythagorean vision of a harmonious universe. A basic element in this concept was the connection between music and number. This crucial discovery is simple enough to be plausibly attributed to Pythagoras directly. Using a monochord, a one-stringed 'instrument," one finds that pleasing pairs of tones are produced only for certain positions of the movable bridge. These harmonious sounds correspond to simple ratios for the lengths of the two parts of the string. Thus the ratio 2:1 produces tones an octave apart; the ratio 3:2 produces a "fifth"; 4:3 a "fourth"; and so on. From these simple observations Pythagoras made the intellectual leap to the idea that the universe is itself in tune—and that number must be the basis for universal harmony, as it is for musical harmony. For Pythagoras pleasant symmetries involving numbers ratified their central role in existence. Consider a single example, the tetractys, which was both number and symbol to the Pythagoreans. Aside from its obvious symmetry,

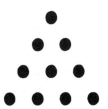

this triangular array contains the number "one" and the One, the Unity, and it contains the mystical perfect number ten, as $1 + 2 + 3 + 4$. The tetractys is symbolic of the cosmos, whose harmonies, however complex, are compounded from numbers.

The cosmologies—there were several—of the Pythagoreans always showed their preoccupation with number. The "music of the spheres," a logical extension from the harmony of vibrating strings, was at the heart of the Pythagorean world. The planets, including the sun and the moon (and in late versions the earth), move in a regular way in a well-tuned world. The tones they make as they rush through the air must therefore be harmonious, in simple ratios 1:2:3:4, etc. What more dramatic evidence of the consonance of all things than these stately motions? The objection that no one hears the celestial music is disposed of simply. We are conceived and born in the presence of these unvarying tones, so we need not, cannot hear them, just as we cannot smell oxygen.[3]

A Pythagorean world model of startling sophistication is described by Philolaus, a contemporary of Socrates. In this Pythagorean universe the earth is not only spherical, but it is moving! As shown in figure 4.2, the earth, the sun and moon, and the planets circle a "central fire" at the center of the universe. The earth goes around the central fire in a diurnal (24-hour) period. The same side of the earth is always toward the central fire, with the Mediterranean on the outside, away from the central fire. This "rolling" motion for the earth is entirely equivalent to the simple rotation of a larger body, and so the Pythagoreans were in

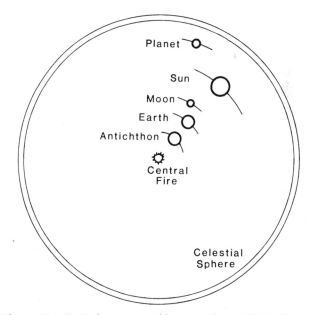

Figure 4.2 The Pythagorean world system advocated by Philolaus.

effect proposing the rotation of the earth.[4] A curious feature of this
cosmology is the "antichthon" or counterearth, a planet that assertedly
remained always between the earth and the central fire. The antichthon
had a dual role. It shielded the earth from the direct rays of the central
fire and, no less importantly for the Pythagoreans, it brought the number of moving objects in the universe to ten—the perfect number. These
ten objects were (in order of increasing distance) the antichthon, Earth,
Moon, Sun, Mercury, Venus, Mars, Jupiter, Saturn, and the sphere of
the fixed stars. The daily rotation of the earth made it unnecessary for
the sphere of fixed stars to move. Characteristically, the Pythagoreans
restored harmony by postulating that the sphere of the stars rotates
imperceptibly slowly, generating the lowest pitch—the fundamental—
in the music of the spheres.

First Science

The earliest philosophies, representing the insights of such wise and
respected men as Thales, were offered as self-evident. The crucial next
step in the development of philosophy and science was the introduction
of rational argument, of logical "proof" for a philosophical position.
This step was taken by Parmenides, of Elea in Italy. Parmenides was
the most influential of several "Eleatics," men who trusted logic above
all, even denying experience when it conflicted with a conclusion arrived at rationally. For Parmenides the whole of Being was indivisible,
unchanging, outside of time. From this premise it followed that change
cannot occur, for "how could being perish?" If our senses tell us that
growth and change, death and decay are real, then our senses must be
untrustworthy. Parmenides' student Zeno framed his famous paradoxes
(arguing that motion is impossible and other such absurdities) in support of the Eleatic position that existence is one, not many. This first
great crisis in philosophy fostered several important attempts to break
out of the logical impasse, including that of the atomists.

The new logical methods fascinated the philosophers. Logical argument based on the assumption that nature is rational, a principal ingredient of any "scientific method," had been discovered. The other element of science, measurement (or experimentation or observation),
would have seemed pointless to the Eleatic. Experience, they would
have said, is illusory. In modern physics one finds a sympathetic echo
of Parmenides' low esteem for superficial appearances. The world of
the quantum is not a commonsense world. Its permanent elements are
not things but equations discovered through logic and "free invention"

(Einstein's phrase)—just as in modern cosmological theories, where, though galaxies vanish, the universe remains, and true knowledge of existence is in the formulas.

With the advent of logic, philosophy began increasingly to be synthetic and philosophers increasingly responsive to the positions of their predecessors. In this way the school of atomism developed, a natural philosophy prefiguring many of the features of Newtonianism. For the atomists, of whom Democritus is best known, the Ionic principle of Necessity had become a mechanical determinism. The real world is made up of solid indestructible "atoms," infinite in number and variety. Motion is the natural state for the atoms, so that when atoms clump together to form objects it is only "until such time as some stronger necessity comes from the surrounding and shakes and scatters them apart."[5]

Planets and suns are formed and destroyed in innumerable vortices throughout an infinite space. Except for the eternal atoms, nothing is permanent. Not even the soul survives death. This sounds quite modern; there is, however, no unbroken line from the atomists to the seventeenth-century natural philosophers. The central physical insight of Democritus was that continued motion is natural. This idea, the principle of "inertia," is Newton's First Law of Motion. Yet the principle of inertia was lost within a century when Epicurus, a later atomist, redefined motion as occurring through a sort of "falling," which is interrupted by the intervention of chance. Epicurus introduced chance to eliminate the oppressive determinism of Democritus. In so doing he also eliminated the coherence that distinguished the atomistic cosmology.

Waiting for the Astronomers

How do things stand now, at the beginning of the fourth century B.C., at the threshold of the golden age of Greek philosophy? There is undeniably a pattern of interaction and progress in the pre-Socratic philosophies, providing ample raw material for the stunning syntheses of Plato and of Aristotle. Similarly, a consensus of a sort can be found with regard to physical cosmologies. In the majority view the universe was spherical and geocentric, and the earth itself a globe. The cosmos was finite but quite substantially bigger than the clam-like Middle Eastern world or the deep-rooted flat earth of Homeric myth. A steady circulation ("the vortex") carried the outer regions and the heavenly bodies around the earth. The cosmos was filled throughout ("the plenum") but not so solidly as the Genesis "waters above the Firmament."

We should not interpret too narrowly any summary such as this one. Explicit descriptions of the world were, before Aristotle, clearly an adjunct to the main order of business, which was to understand the human condition. Every serious thinker offered conjectures, but the world picture was at best consistent with philosophical precepts and at worst capricious—Xenophanes of Colophon is said to have asserted that "there are many suns and moons according to region."[6] Democritus and the atomists are exceptions to this and most generalizations about the early philosophers, in that the deterministic world follows necessarily from the atomistic philosophy; for the atomists it is the main body of philosophy—ethics, aesthetics—that seems arbitrary.

The philosophers were wrestling with a fundamental tension in human existence: the conflict between the disorderly but familiar apparent world and the steadfast logic of the truth which undergirds the world. In their emphasis on abstract symmetry, they had, paradoxically, found the key to eventually understanding the material world. By the fourth century B.C. not only was a spherical earth accepted by educated men, but also it was widely held that the planets move around the earth in circular paths. The idea of circular orbits, an idea traceable to the Pythagoreans, came at about this time to have an unreasonable hold on the minds of astronomers and philosophers alike. The story of the development of science is intimately tied to this theme of circularity.

A different and lasting use of symmetry is illustrated by Anaximander's argument that the earth stays at the center of the universe because there would be no preferred direction for it to move in. This reasoning is an example of the "principle of sufficient reason," which is at the heart of scientific thought. Giorgio de Santillana characterizes Anaximander's system as being "as much an innovation on the way of thinking that came before as the whole of science has been since."[7]

As we shall see, Aristotle incorporated much of the majority view into his cosmology. His genius was that he was able to make the picture plausible, the mechanism reasonable, and the whole consistent with a comprehensive philosophy. It was on this basis, universality and consistency, that Aristotle's cosmological views came to be dominant. Detailed agreement with experience was not yet required of a natural philosophy. Bertrand Russell has written of the golden age of Greek philosophy, "no philosopher seems to have doubted that a complete metaphysics and cosmology could be established by a combination of much reasoning and some observation."

CHAPTER 5

Synthesis

Plato, Aristotle, and Natural Philosophy

It is no accident that the early period in Greek philosophy is known as pre-Socratic, for in a sense the Western philosophical tradition begins with Socrates. We owe much of our knowledge of him to Plato, who was his student, and indeed it is difficult to separate Socrates' ideas from those of Plato. For this reason, and because of Socrates' distaste for the physical world, we shall not say more of him here. Plato, to be sure, shared Socrates' suspicion of physical reality, but his influence on the early Christian and medieval world view was profound. No understanding is possible of the way the Greek view of the universe evolved and was transmitted to late medieval Europe without an examination of Plato's thinking and in particular his view of reality. Plato's successor, Aristotle, if not his philosophical equal, even more profoundly influenced the course of science. He provided what might be called the first great synthesis of human knowledge, and his physics and metaphysics would dominate until the time of Galileo. One of our goals will be to understand Aristotle's cosmology, which persisted for two thousand years. To do so will require that we discuss Aristotle's physics. We shall be called upon to deal with astronomy, especially with the planets, in more detail than heretofore has been necessary. We shall see that his cosmology is grounded in a physics that is both logical and consistent and appeals to common sense. More importantly, the system is an integral part of Aristotle's comprehensive philosophy so that to reject Aristotle's science is to reject some of the most fundamental of his insights. We might contrast this with Plato's attitude, which was that a detailed description of nature, if given at all, was to be taken only as "likely." almost as mere opinion, and certainly not as "true."

Plato had brought to maturity the great traditions in Greek philosophy. In his dialogues Pythagorean mysticism is fused with the unflinch-

ing rationalism of the Eleatics. Inquiry by argument, the technique that
Socrates had perfected, is made even more persuasive by Plato's superb
literary style. It is therefore of great importance that Plato rejected the
study of nature for its own sake. For him true reality resides in the
eternal, and the eternal is in ideas, not in substances. But Plato's ideal-
ism is tempered; although the eternal ideas preexist, human beings must
discover them through reasoning, aided by the intelligence of the eternal soul. Philosophy thus consists in using the mind to search for true
knowledge. Wrong opinion is weeded out, and the soul is purified
thereby. The evidence of the senses must be used in search of truth,
but the substantial is only secondary, an imperfect copy of an ideal
world that exists outside space and time. As a result of this attitude
Plato thought physics and biology to be beneath consideration. Even
astronomy was regarded as an inferior pursuit, despite the assumption
that the motions of the heavens are as close to eternal as anything perceptible can be.

Platonic emphasis on the immortality of the soul fostered an un-
worldliness that regarded life itself as an imposition on the soul, and
neo-Platonic interpretations that ignored Plato's reasoned optimism were
dominant in Christian theology throughout much of European history.
On the other hand the Pythagorean elements in Plato encourage the
hope of discovering a mathematical basis for reality. From this point of
view careful, even quantitative, observation can be an adjunct to mysticism. If measurement and calculation should require a lifetime, mystical contemplation might then be deferred. The position of Johannes
Kepler was close to this, as we shall see in chapter 8. Plato himself
insisted, therefore, that his pupils learn mathematics. Eudoxus (408–355
B.C.), who studied with Plato, provided the first detailed mathematical
model for the motions of the planets. The "homocentric spheres" of
Eudoxus were later adapted by Aristotle as part of his cosmology (discussed later).

Plato offered no fully serious cosmology. His most elaborate discussion of the physical world is in the *Timaeus,*[1] where he describes the
universe as a single, living being fashioned by God on the pattern of
the ideal. The premise of an organic, as opposed to a mechanistic, world
is one that Empedocles had earlier developed. A grand cosmological al-
legory, the *Timaeus* almost by accident had a great influence in the Middle Ages, when the lost Greek traditions were being recovered. By no
means was the *Timaeus* intended to be a detailed description of the universe, since Plato believed that nothing more than a "likeness" could
be achieved in any case.

Aristotle's Physics

Except for Democritus no mature thinker before Aristotle accepted the material world as fully real. From his early youth in Thrace, however, Aristotle was a naturalist who never lost his love of observation, and this permeates his approach to philosophy. Far from ignoring astronomy and physics, Aristotle wrote extensively on each. His universe is coherent and convincing, logically argued, but not always based on premises plausible to the twentieth-century reader. Many of the elements of the cosmologies of the pre-Socratics are incorporated in one way or another. The earth is at the center of the universe, and the celestial region plays the role of the Whirl from which originates all motion. Yet this world is not the poorly defined vortex of Anaximander, nor are the motions that occur dismissed with a vague reference to what is natural and proper. At the heart of Aristotle's description is the division of the universe into two regions, prompting T. S. Kuhn to call it the "two-sphere universe."[2] The two spheres are the earthly or terrestrial region, the region of change and decay, and outside it the perfect, immutable celestial region. The laws that govern the two spheres are different, and there are two kinds of matter, celestial ("ethereal") and terrestrial.

The eternal changelessness of the celestial region allows a reasonably straightforward description. One need only to find a mechanism for the motions in the heavens. Aristotle chose the construction that Eudoxus had proposed, a system of nested spheres that carried the planets and stars along in motions that qualitatively matched the observed motions. The properties of the ethereal matter were those of the ideal: it is "eternal, suffers neither growth nor diminution, but is ageless, unalterable, and impassive."[3]

Within the earthly region, which extended to the surface of the innermost celestial sphere, the one carrying the moon, matter was composed of mixtures of the four elements of Empedocles: earth, water, air, and fire. Appearance and other sensible properties were determined by the proportions of the four elements. Hence transformation corresponded to variations in the elemental composition of an object. Aristotle's "elements" lack much similarity to our present chemical elements, which are unchangeable and are added together to form chemical compounds with various properties. On the contrary, the Greek elements might better be thought of as "qualities." Thus "fire" is present in whatever tends to rise or is brilliant—not only flame but evaporating water, clouds, and bright pigments. "Earth" is that which tends most

strongly toward the center of the universe. "Water" also has a downward tendency but less strongly so than does earth, while "air" tends
upward, or away from the center, less strongly than fire. The elements
displace each other in their "natural" motions. Their natural motions
are along straight lines inward toward or outward from the center of
the universe. Since each element has its own natural motion, transfor-
mations will normally be accompanied by motion. Since there are infinitely many combinations of the elements, the color, density, and other
characteristics of an object become secondary qualities, and transfor-
mations are predictable only broadly. Thus if earth predominates in an
object, it tends to be heavy, cold, and dark. A rock and a piece of metal
are similar in these respects. The transformation of a piece of coal into
a rock-like cinder is understandable; the water, air, and fire have mostly
been driven off in burning.

Aristotle thought that the terrestrial region, where the four elements
are mixed, came to its present state of motion and its complicated configuration through the transmission of motion from the celestial spheres;
the shape of the earthly region, for example, is consistent with the assumed properties of the elements. The earth itself, being mostly earth,
has collected around the center of the universe. This further accounts
for the spherical shape of the earth; the material has traveled from all
directions to the center. As processes occur that alter the relative amounts
of the elements in material objects, motions must occur. For example,
newly liberated fire must rise. The condensation of water must fall as
rain. But the system is not a simply mechanical one, as we shall see.

An important concept in Aristotelian physics is *place*. We see already,
in the tendency of the elements to move, that things tend to go to their
natural place. The agent that causes natural motion is innate. The universe would presumably come to rest if perfect order were achieved—
earth in the center, surrounded by water, air, and fire. But this would
require the heavens to cease their motions. Further, a universe at rest
could not be the abode of man, and Aristotle's philosophy assumes an
ultimate goodness in nature (and in man). The motions generated in the
heavenly regions are purposeful; they sustain life. On the other hand,
the natural philosopher can bring little order to the complexity of terrestrial motion.

The teleological principle in Aristotle's physics is made explicit in his
assignment of the "cause" for motion. In the Newtonian language that
we all normally use, the physical cause for motion is an applied *force*, a
push or pull. Objects act on *each other* to cause motion. An object falls,
we say, because it is pulled downward by the earth's attraction. (In
Aristotelian language, it is driven downward by its natural tendency.)

Aristotle assigns as the *efficient cause* what we have here described as the cause. Aristotle also identifies other "causes" of motion; the most important of these is the *ultimate cause,* which is the purpose or goal for motion. Water, for example, is mixed with fire through the action of sunlight and rises, but the essence of water becomes dominant, a cloud collects, and rain falls. Thus more fundamentally, the efficient cause for rain is the natural motion of water through air, the ultimate cause is that it is *good* for rain to fall in this way. In discussing human events, we still assign ultimate causes, but only for the action of living things. If you should jump out of a window, you would fall because gravity acts on you, but the act itself would be accomplished for some reason. We would agree that the reason for jumping is, however, quite unrelated to gravity and contributes nothing to the fall. A brick dislodged from a wall by the wind has no ultimate cause, in the modern view.

Aristotle assumed space to be completely filled: a *plenum;* no void (vacuum) could exist. To speak of "empty space" was for Aristotle a contradiction in terms; space ("place") is where something is. The *horror vacui* was argued both on theoretical grounds and by appeal to common experience; no one had ever experienced a vacuum. The plenum represents an explicit denial of the atomistic universe of Leucippus and Democritus, who had provided for the motion of the immutable atoms through the void.

Aristotle's objects had no inertia; unless acted on continuously, they stopped moving. Indeed, his objects had no mass in the sense that mass is what gives things weight and allows them to fall. Falling was a natural motion for some objects, but rising was equally natural for others. It might seem that the ever-turning heavenly spheres had inertia, since their motions never stopped. Not so; even in the ethereal region each motion was maintained through the constant intervention of an "unmoved mover." Since the motions of the planets are complicated, a large number of unmoved movers is required—55 is the number given in Aristotle's *On the Heavens.*

To Aristotle terrestrial motion is either "natural," as we have just explained, or violent, as when a rock is thrown upward. The agents that initiate unnatural motion are readily recognized—a blow, a tug at a string. The persistence of unnatural motion is harder to explain, since continued motion is not natural. Aristotle postulated a propagation of the violent action through the medium, the disturbance finally dissipating so that the object stops. Thus when an arrow is shot, the impelling disturbance is not confined to the bow string but continues through the air; the air pushes the arrow along.

In the absence of an understanding of inertia there was no chance for

Aristotle to obtain a correct quantitative description of motion. In those few instances when he proposed formulas for predicting motion, the relationships are grossly wrong. A contributing difficulty was that there was no way of measuring short time intervals, so that judgments about velocity and acceleration were difficult or impossible. Consequently what we would call the variables of motions were defined vaguely if at all, the emphasis being on the accomplishment of a distance in a time interval (that is, on average velocity). Aristotle said that a heavy object seeks the center in proportion to its size, and this is qualitatively correct: If a heavy ball and a lighter one of the same size are dropped at the same time, the heavier ball will arrive somewhat sooner, although it would not do so in a vacuum. But the transit time was not subjected to measurement, and by "proportion" Aristotle did not intend "ratio." In any event, because of the dominant role of ultimate cause, there was little incentive to make quantitative measurements. The future is determined not fundamentally by the motions of things but by the ultimate goodness of nature. Motion assumes its full importance in a mechanics that deals only with efficient causes.

The Homocentric Spheres

Astronomy, especially planetary astronomy, had begun to emerge as a science during the fourth century B.C. This was due in considerable part to the development of geometry. Plane geometry was by Aristotle's time in a form much like that taught in schools today, and many of the significant problems in spherical geometry had been solved. Plato recognized that the motions of the planets required explaining; it was no longer enough to assert that the planets move in various circles around the earth. Eudoxus, who was a master of geometry and who presumably had obtained in Egypt copies of astronomical records, set himself the task of reconciling the observed wanderings of the planets with their assumed perfection.

Before undertaking to explain Eudoxus' results, let us review some of the elementary features of the motions in the sky from a geocentric point of view, that is, as they actually appear to us. From nearest the earth to most distant, the seven "planets" known in antiquity (in the order most often given) were Moon, Sun, Mercury, Venus, Mars, Jupiter, and Saturn.[4] If these seven objects are temporarily removed from the sky, the only objects seen are the stars, forming a fixed pattern on the celestial sphere, which spins as a whole from east to west about the poles within a period of 23 hours and 56 minutes, called the sidereal day. (We here ignore precession of the equinoxes.) If we now replace all of the planets except the sun, they will be seen to share the east–west motion of the celestial sphere but will generally move more slowly than the stars, rising on the average somewhat less frequently than once every sidereal day. This is most pronounced for the moon, which falls behind daily by about 13° (or 50 minutes by the clock). Saturn, which is supposed to be closest to the celestial sphere, loses only about 1/30° per day. As seen against the stars, the planets also wander north to south and back; all of them, however, remain near the ecliptic, the track of the sun through the zodiac. This implies that the orbits of the seven planets all lie near a single plane. The brightness of each planet varies as it moves, Mars showing the most dramatic changes. It was apparent that variable brightness could indicate either varying distance or changing luminosity. Finally, we replace the sun, so that the universe is again complete. The angular motion of the sun is not unlike that of the other planets: slower than the stars in its westerly motion and wandering north to south through the zodiac. The unique brightness of the Sun means that its position among the stars cannot be determined directly but must be inferred from observations made at night.

The motions of the moon and the sun are not quite uniform, but both move inexorably eastward relative to the stars. The other planets describe periodic westward swings called retrogradations. For Venus and Mercury the retrograde motions are dominant, taking the planets westward past the advancing sun, from the evening sky into the morning sky and back again. For Mars, Jupiter, and Saturn the retrograde loops are not so wide and the correlations with the sun are more subtle. Retrograde motion, although not so spectacular for the latter planets as for Venus and Mercury, is far from being an obscure phenomenon. It is in fact among the most striking motions in the heavens, and Eudoxus was the first to account for it mathematically.

In a heliocentric model, this perplexing motion has a simple explanation. As figure 5.1 illustrates, every time the earth overtakes or is overtaken by another planet, retrograde motion is observed. But what, in a geocentric universe, explains the strange willingness of a planet to stop (relative to the stars) and retrace a part of its path? We have already seen that circular motion was thought by most Greek philosophers to be natural for heavenly bodies. Retrogradation, looping back, required compounded circular motion, circles carried by circles. The problem of reproducing the path of a planet then became that of building up a complicated motion from circular constituents. Of course, any reasonable curve can be produced by combining sufficiently many simple curves. This technique will emerge again when we examine the Ptolemaic system in chapter 6.

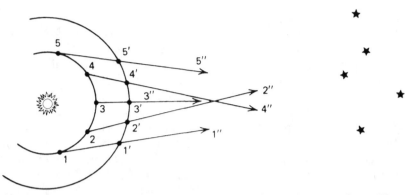

Figure 5.1 Retrograde motion is easily explained in a heliocentric cosmology. When the earth overtakes an outer planet (Mars, for example), its successive positions against the background stars are as shown. The usual eastward motion of the planet is interrupted for a while, as the earth overtakes it ("opposition") and the planet reverses its course and moves back westward. Eventually, as the earth leaves the planet behind, the latter resumes its eastward motion in the sky. From W. H. Jefferys and R. R. Robbins, *Discovering Astronomy* (New York: Wiley, 1981).

Eudoxus' model treated the motion of each of the seven planets independently, but in each case a variation of the same geometrical scheme was used. The planet is embedded, jewel-like, at the equator of the innermost sphere of a set of nested, spinning spheres. Each sphere turns about its axis at a constant rate while its axis moves with the next sphere. The motion produced is not uniform but repetitive and compounded of circles.

As Eudoxus' system has been reconstructed (fig. 5.2), largely from the commentary of Simplicius in the sixth century,[5] each planet is moved by one of two combinations of spheres: three for the moon or sun, four for each of the other planets. In each instance the system begins with two outer spheres, the outermost having a north–south axis and the second tilted by 23° to the first. These provide for the daily turning of the celestial sphere and for an average path eastward along the ecliptic (here defined as "the circle through the zodiac," rather than as the sun's path). For the moon and the sun the system is completed by a third sphere, slightly tilted from the second so that the object is periodically carried away from the ecliptic. This is necessary for the moon, whose deviations of as much as 5° from the ecliptic had long been recognized as being related to the eclipse cycle. The third sphere for the sun is without justification since the sun does not wander from its circle.[6]

For each of the five remaining planets Eudoxus added two spheres to the basic pair described above, for a total of four. His use of only two spheres to produce both retrograde motion and deviation from the

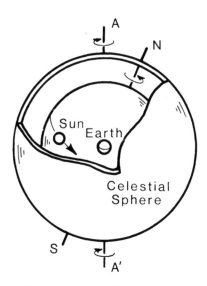

Figure 5.2 The use of two homocentric spheres to reproduce the observed motion of the sun. The outer sphere moves like the celestial sphere, once each sidereal day (23^h 56^m). The axes of the inner sphere are attached to the outer sphere at points A and A', and so the inner sphere, in which the sun is fixed, rotates slowly in the opposite sense. Thus (1) the solar day is longer (24^h) than the sidereal day, and (2) the sun moves through the stars annually.

ecliptic was a brilliant piece of mathematics. He recognized that a point on the equator of the inner of two spheres traces out a figure eight if the two spheres rotate at equal rates in nearly opposite directions (see fig. 5.3). This curve is traced out once for each rotation of the two spheres and with a size determined by the orientation of the two axes of rotation. The effect of compounding the figure eight with the motion of the two outer spheres is to produce something that at least resembles the actual retrograde motions of the slower planets Jupiter and Saturn. For Mars and swift Venus the four-sphere model achieves only retarded motion, not actual retrogradation; for Mercury it is not clear what orbital parameters Eudoxus used. Thus despite their elegance the nested spheres are a qualified failure.

By the time Aristotle incorporated the homocentric spheres model into his physics of the heavens, it had been further modified by Callippus, who added a total of six spheres to Eudoxus' 27 [(3 × 2) + (5 × 4) + 1 (for the actual celestial sphere)] in order to improve the agreement with observation, an approach so much in keeping with the spirit of modern theoretical science that J. L. E. Dreyer has described it as giving astronomy its true start.[7] As we pointed out earlier, a system of arbitrarily many-nested spheres could produce a path of any shape whatsover. However, there are difficulties even with five spheres for the inner planets.[8] And whatever its path, in the homocentric sphere model the planet is at a fixed distance from the center of the earth, so that no variation in brightness is predicted. For all its symmetry and geometrical subtlety, the work of Eudoxus survived largely because of the historical accident that it was the best model available to Aristotle.

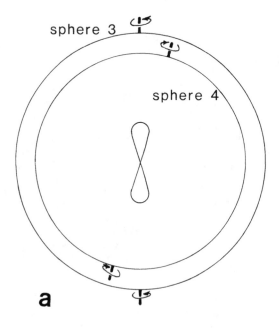

sphere 3

sphere 4

a

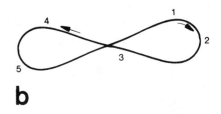

b

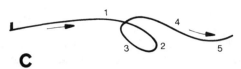

c

Figure 5.3 Two homocentric spheres with equal and almost opposite angular velocity produce a motion for a point on the equator of the inner sphere which is as shown in (a) and (b). The "spherical lemniscate," or "hippopede" (so called because horses were exercised on a figure-eight path) produces retrograde motion (c) when added to a uniform motion along the direction of its axis.

As geometricians and astronomers rather than "physicists," Eudoxus and Callippus were interested in showing that the irregular motions of the planets were accessible to the mathematics of circular motion. The homocentric spheres model became a machine only after Aristotle incorporated them into his cosmology. He, on the other hand, was less interested in the geometrical details than in the necessary attributes of the heavens. If there are spheres, they must be made from ether; hence the spheres themselves are crystalline, impenetrable, eternal. The motions in the ether, whatever astronomers find them to be, continue forever and are caused by the prime movers.

Moreover, all terrestrial motions are traceable to the motions transmitted by friction from the innermost sphere, that of the moon. In this sense all motion is from the unmoved movers. Aristotle increased the number of spheres from Callippus' 33 to 55 so as to make the celestial regions a part of the plenum. Since there could be no empty space, the spheres of one planet must abut against those of the neighboring planets. But, Aristotle felt, great friction would result from the different speed of the spheres. So he insulated each of Eudoxus' sets of spheres with additional counterrotating spheres that successively canceled their motions. The result of the added spheres was that the innermost sphere of each planetary set rotated once each sidereal day. Since the outermost sphere of each set already moved in this way, there was no slipping between sets of spheres. The moon alone was left bare. If friction would occur between ethereal spheres belonging to adjacent planets, how could friction be absent among the spheres moving a single planet? These and like questions would have been within the province of later natural philosophers, but there was to be no real successor to Aristotle among the Greeks.

CHAPTER 6

Orb in Orb

It is often said that, despite important progress in mathematics and astronomy, Greek science on the whole declined during the Hellenistic (c.323–146 B.C.) and Greco-Roman (146 B.C.–A.D. 476) periods. In many respects it was more a shift in focus, to a greater preoccupation, following Plato's lead, with *saving the phenomena* in astronomy, rather than with the dynamics of the universe. To some this is seen as a retreat from rational speculation about the world that exposed a declining faith in the ability of reason to uncover the secrets of the universe, an acceptance of Plato's belief that at best our descriptions of the world will be only an approximate likeness. Though this may be so, the argument has to deal with the giant figures of Aristarchus, Archimedes, and Hipparchus, among others, whose efforts in the third and second centuries B.C. to understand nature represent the highest achievements of Greek science.[1]

Post-Aristotelian Greece

Hipparchus is a transitional figure, author of a lost work on falling bodies but often content with purely mathematical astronomy. Of all the scientific figures of antiquity, Archimedes is no doubt the greatest; it would be the fifteenth century before Western science could boast that it had returned to the standards of mathematical and scientific investigation he represents. Yet the accomplishments of the era were mainly in mathematics (Euclid, Apollonius) and astronomy (Hipparchus, Ptolemy), and increasingly the goal in the latter science was to save the appearances. This was in part because Plato and Aristotle had been so successful in addressing the fundamental problems of ethics, metaphysics, and epistomology and was also a consequence of the subsidiary role assigned by Plato to the physical world.

But Greek philosophical thought (in the modern sense of the term) had far outstripped progress in physics and had provided a description of the world and of motion that inhibited empirical discovery of the correct laws. The authority of Aristotle in these matters was almost unchallenged, and in a geocentric model, no progress in understanding how the planets move, for example, was possible. The reception of Aristarchus' heliocentric ideas is evidence that the time was not ripe for serious consideration of the mobility of the earth—and yet within 400 years after recovery of the work of Aristotle, Archimedes, and Aristarchus, not only was the heliocentric theory accepted by many astronomers, but also much progress had been made toward a mature science of mechanics. It would seem then that the *authority* of Plato and Aristotle, their division of the universe into a terrestrial and a celestial sphere, the concepts of natural and violent motion, and the Platonic assertion that truth was not to be found in the natural world provide the explanation for the "decline" in Greek science.

Platonic and Aristotelian metaphysics were to dominate Western thought for nearly two millennia, and Aristotle's physics would not be substantially questioned until the fourteenth century. In astronomy the turn from rational speculation about the universe, which we have traced from Anaximander to Aristotle to the merely predictive and computational became more and more pronounced.

Interest in dynamics, the full explanation of motion, had never been widespread, and by the end of the first century B.C. the satisfactions of accurate observation and prediction had resulted in a near abandonment of celestial dynamics. Moreover, the "otherness" of the celestial regions (the "two-sphere" mentality) was accepted by philosophers and astronomers alike. As a result no one, not even the great scientist and engineer Archimedes (c.287–212 B.C.), could see any direct relationship between astronomy and terrestrial physics. The remarkable advances Archimedes made in the physics of stationary objects[2] ("statics" and "hydrostatics") consequently led nowhere. The separation of astronomy from physics closed, for a time, the only avenue for explaining the visible world.

The achievements of the Greek astronomers are nevertheless remarkable. Already in the fourth century B.C., Heraclides of Pontus had asserted that the stars are fixed and that the earth rotates on its axis. He also resolved one of the puzzles of the planetary motions by proposing that Mercury and Venus revolve about the sun, an idea resurrected nineteen centuries later by Tycho Brahe. Heraclides proposed, as did Apollonius independently, a successful alternative to the planetary kinematics of Eudoxus, Callippus, and Aristotle. The alternative was one in which the motions of the planets were explained in terms of epicycles and eccentrics. This method, which came to be called the "Ptolemaic system" is discussed in detail later.

Aristarchus of Samos (c. 310–230 B.C.) proposed for the first time a fully heliocentric universe in which the earth and the other planets revolved about the sun. Unfortunately only brief, second-hand reports of Aristarchus' mature work are preserved, and so it is not possible to know how fully he may have developed the heliocentric system. Probably he recognized that simplifications in the actual motions of the planets could follow if the earth were allowed to move. He realized that his proposal required the distance to the fixed stars to be vastly greater than was commonly supposed; otherwise the stellar patterns would shift as the earth in its annual motion approached and receded from a given star.

Our best account of what Aristarchus seems to have taught is given by Archimedes in his *The Sand Reckoner:* [3]

You are aware that universe (*cosmos*) is the name given by most astronomers to the sphere whose center is the Earth. This is the common account as you have heard from astronomers. But Aristarchos of Samos brought out a book consisting of some hypotheses, wherein it appears, as a consequence of assumptions made, that the [real] universe is many times greater than the one just mentioned. His hypotheses are that the fixed stars and the Sun remain unmoved, that the Earth revolves about the Sun in the circumference of a circle, the Sun lying in the middle of the orbit, and that the sphere of the fixed stars, situated about the same center as the Sun, is so great that the circle in which he supposes the Earth to revolve bears such a proportion to the distance of the fixed stars as the center of the sphere bears to its surface.

It is unlikely that Aristarchus made any detailed calculations based on the heliocentric system. In any event Aristarchus' daring suggestion failed to gain ascendancy, and for seemingly sound reasons. It was, for example, difficult to understand why clouds and living things should not be swept off a moving earth. More to the point, the geocentric universe is an anthropocentric one. Even given the Greek spirit of free inquiry, it took great intellectual courage for Aristarchus to so dethrone man. He was accused of impiety for asserting that the earth moves, though in the third century B.C. such a charge lacked force. His idea, which was to influence Copernicus, was forgotten during the intervening centuries. Of the heliocentric system Pannekoek says, "It did not, so far as we know, present itself as an explanation of the irregularities in the planetary motions; it remained a bold, ingenious, but isolated idea."[4]

Aristarchus was also the first to make a direct measurement of the relative distances to the sun and the moon.[5] Although the value he obtained for the ratio of these distances was not close to the correct one, the method was entirely valid. Aristarchus measured (or estimated) the angle between the moon and the sun at first or third quarter ("quadrature") and obtained a value of 87°. The correct value is much closer to 90° (89°51') and the final answer is very sensitive to the precision of the measurement of this angle.

After about 250 B.C. the intellectual center of the Hellenic world shifted to Alexandria. Although such scholars as Archimedes and Hipparchus (c.190–130 B.C.) worked elsewhere, the Alexandrian museum, with its great library, became the center of Greek science. Hipparchus was a transitional figure who wrote on problems of dynamics but is now known to us principally as the greatest astronomer of antiquity. A renowned calculator, he obtained a value for the lunar month accurate to within a second, although this result was largely borrowed from the Babylonians of the Neo-Babylonian and Persian periods. He was also able to distinguish between the tropical and sidereal years, a feat that may have followed from his discovery of precession by the appearance of a new star, or nova,[6] in 134 B.C. His interest turning to the "constant" heavens, Hipparchus compiled a star catalogue in which some 850 stars were divided among six magnitudes of brightness. Comparison with earlier catalogues revealed that the heavens were shifting, and Hipparchus correctly recognized that precession would account for the changes. The phenomenon is an ingredient of many ancient myths (see chapter 2), but Hipparchus through observation removed precession from the realm of superstition.

The Almagest and the Ptolemaic System

Claudius Ptolemaeus, or Ptolemy, (85–165) brought Greek observational astronomy and the mathematical techniques of astronomy to their highest development. As we have suggested, by the time of Ptolemy Greek theoretical cosmology and astronomy had largely degenerated into a collection of computational techniques. Questions of the mechanical workings or dynamics of the planetary system had been dropped, while observational and predictive astronomy had reached new heights. Ptolemy is known principally for his astronomical masterpiece *The Almagest* and for his elaboration therein of the kinematic model of deferents and epicycles, from which the tables of *The Almagest* were computed. The Greek title of the work was *Mathematical Syntaxis,* and it was known to the Arabs as *Al Magisti,* "The Great," hence *"Almagest."*

Although a complete description of the Ptolemaic system is beyond the scope of this work, we shall illustrate most of the salient features of

the planetary model that so successfully served the needs of Western
man until the seventeenth century. As noted in chapter 5, the principal
problems confronting the Greek astronomers in constructing a model
of planetary kinematics were (1) the variable velocity of the planets, the
"anomalistic motion"; (2) the problem of stationary points and retrograde motion; (3) the varying distances of the planets from the earth,
exhibited in the changing size of the moon and changing brightness of
the planets. It proved to be possible to approximately solve these problems, which are not entirely independent, by the device of the epicycle
(fig. 6.1A). An epicycle is a circle that carries the planet, while the
center of the epicycle itself revolves about the earth in a circular path
called the deferent.

   An alternate and partially equivalent method involved the use of an
eccentric: the planet moves in a circular orbit about a center displaced
some distance from the earth (fig. 6.1B). Ptolemy, of course, did not

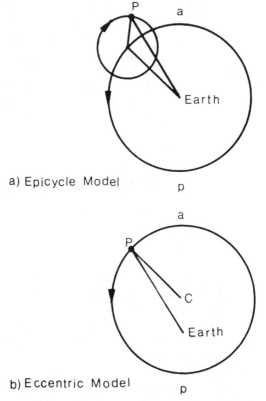

a) Epicycle Model

b) Eccentric Model

Figure 6.1   (a) The epicyclic model. The epicycle moves on a circle (deferent) with center
at O the earth. (b) The eccentric model, with the earth displaced from the center.

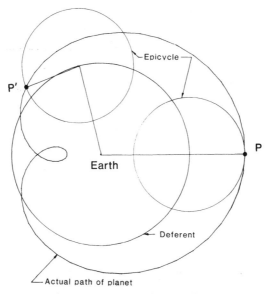

Figure 6.2 An example of retrograde motion produced by an epicyclic model in which the angular velocity of the planet about the center of the epicycle is twice that of the center of the epicycle about the earth, and in the same direction.

invent these methods, but in his work they became precise, quantitative techniques for computing planetary longitudes. Epicycles and eccentrics are obviously consistent with the theme of uniform circular motion as the only natural motion in the heavens. The overall motion of a planet on an epicycle is not circular, but it is "compounded circular." The eccentric provides an actual circular path but is unsymmetric in that the earth is not the center. This objection can be met in good Pythagorean fashion by observing that the center of the eccentric can be thought of as moving around the earth with an indefinitely long period, so that the planetary orbit has the earth as its logical center. From this point of view epicycles and eccentrics are variations on a single method. And indeed the mathematical equivalence of the two descriptions had been shown by Hipparchus. Depending on the rate of revolution of the epicycle on the deferent and the rate of the epicycle, a planet can be made to execute any desired number of loops; for example, in the case in which the epicycle rotates (directly) twice for every revolution of the deferent, we obtain the result shown in fig. 6.2.

The problem of the sun is easiest of all, with nothing more required than a single eccentric or the equivalent epicycle, with the earth at the center. By adjusting the appropriate parameters, it is possible to make the period from the spring equinox to the autumnal equinox six days

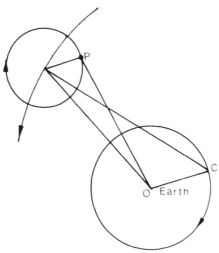

Figure 6.3 Ptolemaic lunar model. The large variation in the speed of the moon in the sky caused by the eccentricity of its orbit required an epicycle which moved in and out with respect to the center O, where the earth was located.

greater than its complement, from autumn to spring. The Ptolemaic lunar theory on the other hand, is much more complicated, principally because the line of apsides[7] of the lunar orbit rotates in space so rapidly, requiring a "crank mechanism" (fig. 6.3), which yields a variable distance between the earth and the epicycle.

Although the outer planets were modeled rather easily, the new device of the *equant* ("punctum equans") had to be introduced. In this case the motion was required to be uniform, not as seen from the earth or from the center of the deferent, but from a point equidistant from the center on the opposite side from the earth (fig. 6.4A). Thus the epicycle no longer revolved with uniform circular motion, violating the sine qua non of all ancient cosmological models. Copernicus was to be especially troubled by the sacrifice of symmetry that this entailed. The inner planets, Mercury and Venus, which never stray far from the sun as seen at the earth, required the most elaborate models of all, a further development on the theme employed for the moon. Not only was a crank mechanism required, but also an equant point was used with respect to which the motion was uniform (fig. 6.4B).

With these devices Ptolemy was able to fit the observed planetary longitude within the observational accuracy of the period, but the problem of latitudes had also to be dealt with.[8] Ptolemy's theory of planetary latitudes was not very satisfactory and moreover seems to have been

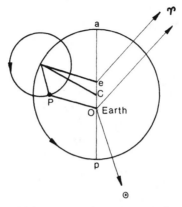

Figure 6.4A Ptolemaic model for an outer planet (Mars, Jupiter, or Saturn). The motion of the center of the epicycle is uniform as seen from the equant *e* rather than from the earth (O) or from the center *C*.

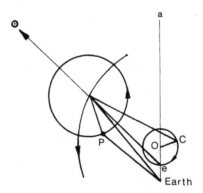

Figure 6.4B Ptolemaic model for the inner planets (Venus and Mercury).

poorly understood by medieval astronomers, but by employing epicycles inclined at appropriate angles to the planes of the deferents he was able to approximate the apparent latitude excursions of the moon and planets.

One cannot complete a discussion of that great paradigm, the Ptolemaic system, without pointing out what might be seen as a curious lapse on the part of the Greek astronomers. The Ptolemaic model for the inner planets in its simplest form, without the "crank-mechanism," consists of an epicycle whose center must always be in the direction of the sun, so that the period of the motion at the center of the epicycle is one year. In the case of the outer planets the line joining the planet with

the center of its epicycle motion has a period of one year. Why should
the sun's period of one year be fundamental in the Ptolemaic model of
every one of the five planets? Whether that question was seriously asked
we do not know, but we might suggest that it ought to have been. But
would the correct answer, that the period of one year is universally
present because it is the earth's period of revolution about the fixed sun,
have survived the hostile reception it would have received and the arguments from Aristotelian physics and metaphysics, not to mention
common sense, that demonstrate the earth does not move? In all probability it would not have, for it was the fate of Aristarchus' heliocentrism to be ignored for 1800 years.

The Almagest is perhaps the greatest single work in the history of
astronomy. Not only is the Ptolemaic system in its most sophisticated
form developed in great detail, and a whole system of calculations described, but tables of the predicted positions of the planets are given
that provided essentially complete agreement with observation. Moreover, it was possible to derive relative distances to the planets. Given a
single absolute distance, the distance to the moon, along with estimates,
crude though they were in the time of Hipparchus and Ptolemy, of the
apparent angular diameters of the planets, it was possible to obtain figures for the absolute dimensions of the planets. The Ptolemaic system,
as embodied in The Almagest, was for the second century a computational achievement of the very highest order. It formed the basis for
some seven centuries of Arab astronomy and provided a secure world
view that was virtually unchallenged until the sixteenth century. Neugebauer has remarked that "one cannot read a single chapter of Copernicus or Kepler without a thorough knowledge of the Almagest." And
yet in the last analysis, Ptolemy was an astronomer, not a physicist; the
question of the physical reality of the model was irrelevant.[9] It sufficed
that it had great predictive value. It might be noted that a number of
commentators, from D'Alembert in the eighteenth century on,[10] have
disputed the originality of data represented by Ptolemy as his own.
Whether further research supports the thesis that he fabricated observations or not, the enormous influence of the Almagest in the history of
astronomy remains secure.

To Save the Appearances

Well before the time of Ptolemy, emphasis in Greek astronomy had
begun to shift from rational speculation about the nature of the universe
to predictive and computational astronomy. With Aristotle, Greek

"physics" reached its heights, and if we would more properly characterize much of Aristotle's physics as "metaphysics," its historical impact was enormous. Moreover these attempts to lay bare the workings of the universe, even though constrained by a priori assumptions about the lunary or celestial world, represent the greatest achievements of Greek speculation about the cosmos.

The homocentric sphere model of Eudoxus and Calippus played a crucial role in the development of astronomy, for their choice had been to "save the appearances" (Plato's phrase), to settle for fitting the observations with the most economical model possible. To the historian of astronomy J.L.E. Dreyer,

Scientific astronomy may really be said to date from Eudoxus and Calippus, as we here for the first time meet that mutual influence of theory and observation on each other which characterizes the development of astronomy from century to century. Eudoxus is the first to go beyond mere philosophical reasoning about the construction of the universe; he is the first to attempt systematically to account for the planetary motions. Philosophical speculation unsupported by steadily pursued observations is from henceforth abandoned.[11]

Dreyer here describes a mature way of doing science, in which theory and observation must agree, even at the price of altering or rejecting the theory. Notice, however, that the "theory" of Eudoxus and Calippus is not a description of nature. A historic turning had occurred for astronomy. The broader questions of natural philosophy began to have less interest for the astronomers.

Aristarchus' heliocentric system offered the last possibility of providing a physically "true" description that might also save the appearances. But there is no way to know the spirit in which Aristarchus made his suggestion. When a century earlier Heraclides had speculated that one could save the appearances by assuming the stars fixed and the earth turning, he had seen it only as a hypothesis equivalent to the geocentric system; and he believed that it was not within the grasp of the astronomer to ascertain the true nature of heavenly motions. Hipparchus showed that the two alternative descriptions, the one using epicycles on deferents, the other employing eccentrics, were formally equivalent, but no attempt was made to distinguish between the two on "physical" grounds.

It was clear to Theon of Smyrna, predecessor of Ptolemy, that "No matter which hypothesis is settled on, the appearances will be saved. For this reason we may dismiss as idle the discussions of the mathematicians, some of whom say that the planets are carried along eccentric circles only, while others claim that they are carried by epicycles, and still others that they move around the same center as the sphere of the fixed stars."[12] Ptolemy said simply, "We must, as best we can, adapt the simplest hypothesis to the heavenly movements. But if these prove insufficient, we must select others that fit better."[13] Indeed Ptolemy was willing to accept a lunar theory that led to enormously incorrect changes in the apparent size of the moon but that gave the longitudes with great precision.

During the Hellenistic and post-Hellenistic eras, these contrasting approaches of the physicist (or metaphysicist) and the astronomer drew a great deal of attention. Following Ptolemy, Proclus said, "these circles exist only in thought; they interchange natural bodies and mathematical concepts; they account for natural movements by means of things which have no existence in nature."[14] Duhem remarks that Proclus' doctrine is not unlike modern positivism, in that a sharp distinction is made between what is accessible to human knowledge and what is not.[15] The constructions used to save the phenomena need not be truthful or even plausible. This view was, of course, not universal, and there were those, who, like Simplicius, adopted a position intermediate between those of Aristotle and the operationalists.

During the early Middle Ages, when astronomy was largely dormant in Latin Europe, Islamic astronomers struggled with the same question. Some attempted to divest the Ptolemaic system of its abstract character by seeking a physical realization of it; they built planetary models with spheres of brass rolling within larger spheres. However much they revered Aristotle, the Islamic scientists were not constrained to reconcile their astronomical models with Aristotelian physics, which is not fully consistent with the Ptolemaic system. Not so fortunate were the scholastics of the European cathedral and monastery schools of the late Middle Ages. Only by the most subtle logic could Ptolemy and Aristotle seem to be united. Thomas Aquinas wrote this justification for adopting the Ptolemaic system while at the same time adhering to Aristotelian metaphysics:

We can account for a thing in two different ways. The first way consists in establishing by a sufficient demonstration that a principle from which the thing follows is correct. Thus, in physics we supply a reason which is sufficient to prove the uniformity of the motion of the heavens. The second way of accounting for a thing consists, not in demonstrating its principles by a sufficient proof, but in showing which effects agree with a principle laid down before hand. . . . Provided he (the astronomer) has the means of correctly determining the places and motions of the planets, he does not inquire whether or not this means that there really are such orbits as he assumes up in the sky; that investigation concerns the physicist.[16]

Yet the question was by no means settled. At the foundation of the Copernican revolution of the sixteenth and seventeenth centuries was the conviction that the phenomena must be saved, but only by means of hypotheses conformable to the principles of physics. Owing to the success of the Newtonian synthesis, scientific introspection was deeded back to the philosophers. So our hypothetical man in the street, the twentieth-century Newtonian, on being asked, "Does physical science describe the real world, or does theory-making result in 'saving the appearances' only?" might respond that the question is ridiculous, that certainly physics describes reality and it saves the appearances at the same time. Many practicing scientists of the present day would agree with this claim of completeness for physical science. But many recent historians and philosophers of science have tended to reassert the narrower view, that science provides only an economical description of observed phenomena and that the concepts used ("atom," "neutron," "force") need not be thought of as real. So the question is an open one; we invite the reader to keep it in mind as we continue to explore cosmological ideas.

CHAPTER 7

Medieval Europe

As we have seen, the "decline" in Greek science was neither abrupt nor universal; at Alexandria astronomy, mathematics, and to a lesser extent mechanics continued to flourish throughout the Greco-Roman period, and the works of Ptolemy, Hero, Diophantus, and Galen sustained the Greek tradition. Yet there were signs that the Greeks were becoming "fatigued with rationalism," and physics was largely abandoned. As the Roman Empire expanded, contacts with the Near Eastern cultures exposed the Greco-Roman world to a variety of oriental religious and mystical movements, which encouraged a preoccupation with revealed truth and mystical insights.

Furthermore the Romans emphasized applied science; this led to remarkable achievements in technology and engineering but to little progress in science and mathematics. Calendar reform is an example of the practical mode of Roman astronomy. The characteristic form of Roman science was the encyclopedia, descriptive and comprehensive but largely derivative. This emphasis continued into the early Middle Ages, when science was based almost entirely on Roman sources. The best-known and most accessible product of Roman science is the fine atomist poem "On the Nature of Things" by Lucretius (Titus Lucretius Carus), who died in 55 B.C. It was to have enormous influence when rediscovered in the fifteenth century, and it helped pave the way for modern atomism.

After about A.D. 180, about the time of the death of Marcus Aurelius, the forces tearing at the integrity of the Roman state began to weaken the empire and over the next two centuries the story was one of internal strife, authoritarianism, and mysticism. The extreme economic inequalities, producing division by rank and economic status into classes whose interests were irreconcilably opposed to each other; the ponderous, self-serving, arbitrary bureaucracy; increasing militarism; suppression; all this made recovery or change impossible. The middle class, in which the strength of Rome lay, gradually disappeared under attacks from the peasants and working class and under the burden of confiscatory taxes. Rigid totalitarianism appeared and worship of the Emperor was demanded.

By 320, Christianity had become the official religion of the empire, and after Rome fell into chaos and social decay late in the fifth century the church alone kept alive the Roman tradition and prevented Europe from relapsing into barbarism. Yet while it preserved some of the results of eight centuries of Greek philosophy and science, there was within the church little genuine interest in science. Natural philosophy was tolerated as a source of illustrations of the truths of morality and religion, at least those teachings judged not to be contrary to the scriptures. Not surprisingly the sophisticated cosmologies of the Greeks soon disappeared from the Christian West. St. Ambrose expressed a typical attitude: "To discuss the nature and position of the Earth does not help us in our hope of life to come."

Throughout the early Middle Ages there could be little doubt about the subsidiary position of science to theology. Moreover the church siphoned off many of those who might have pursued natural philosophy but who now were preoccupied with Christian dogma. St. Augustine (354–430) was the principal channel through which the traditions of Greek thought passed into Latin Christianity and was chiefly responsible for the Platonic character of early Christian theology. Augustine was greatly influenced by the great pagan Neo-Platonist Plotinus. He acknowledged the role of reason in bringing the unbeliever to faith and argued that we come to have knowledge of the eternal through sensory experience and reason coupled with the unchanging, eternal Being, God. Yet it was not until Thomas Aquinas in the thirteenth century that the unity of truth, whether obtained through reason or revelation, was fully acknowledged.

Sixth to Fourteenth Centuries

The scientific inheritance of the Latin West was preserved largely in the compilations of the Latin encyclopedists. Until the twelfth century, when translations of Greek and Arab works began to come into western Europe, Pliny's *Natural History* [1] was the principal source of facts about the natural world, even though it was weak in its understanding and explanation of the Greek sources and emphasized the odd and curious. Another compiler whose work helped to keep alive the scientific learning of the Greeks was Isidore of Seville (560–636), whose

Etymologies[2] was a source of knowledge of all kinds for centuries. The earth, he thought, was shaped like a wheel, encircled by the ocean. Around the earth were the concentric spheres of the planets and stars, and outside the last sphere was heaven, the abode of the blessed. Severianus, Bishop of Gabula, in his *Six Orations on the Creation of the World,*[3] endeavored to explain the crude cosmogony of Genesis. This very primitive picture of the world as flat, in the shape of a tabernacle, was generally accepted by early Christian writers.

Earlier, in the fourth century, Lactantius had heaped ridicule on the idea of the spherical earth.[4] Ambrose and his disciple Augustine held more moderate views, accepting a spherical universe, but were nonetheless committed to the waters above the firmament described in Genesis. This anomalous feature of the cosmology of Genesis was forced on the Christian writers in spite of its clear incompatibility with Greek philosophy. Origen (c.185–254) is supposed to have interpreted this doctrine allegorically, and John Philoponus, the sixth-century Neo-Platonist philosopher and theologian, simply argued that the writers of the Old Testament were unconcerned with technical questions of astronomy and cosmology. Nonetheless, the rejection of the sphericity of the earth as a pagan idea, or reliance on the tabernacle image of the scriptures, was widespread. Some went so far as to deny even geocentrism—the earth is so heavy that it is at the *bottom* of the universe! Most of what the Middle Ages knew of logic was due to the translations of Boethius (c.480–524), while the only source of Plato's ideas was a translation of the *Timaeus,* an untypical work, by Chalcidius.

Philoponus was one of the first scholars to criticize Aristotelian mechanics.[5] One of the Aristotelian arguments against the possibility of a void was that the speed of an object is determined by the ratio of impelling force to resistance, so that in the absence of any resisting medium the speed would become indefinitely large. Philoponus asserted that the speed would be proportional to the *difference* between force and resistance; as a description of motion this idea is wrong, but the assignment of a subsidiary role to air resistance was helpful. Thus Philoponus was led to reject the Aristotelian idea that a projectile, after losing contact with the projector, is moved along by the turbulence of the air. He assumed the air to play an insignificant role and introduced a "kinetic force" that is initially imparted to the projectile and that sustains its motion. The concept of kinetic force had considerable historical influence, particularly on Islamic philosophers, and was very close to the late medieval concept of *impetus.* Philoponus is also credited with having foreseen the result of the famous experiment with falling spheres that Galileo was supposed to have carried out at Pisa eleven centuries later.

Prominent among those who in the darkest part of the Middle Ages kept alive the intellectual traditions of the past was the Venerable Bede (673–735), who knew Pliny's world view and who derived much of his material from Isidore. Bede held that the earth was spherical, surrounded by seven heavens of air, aether, Olympus, fiery space, the firmament of heavenly bodies, the heaven of the angels, and the heaven of the Trinity. He showed how to use the 19-year (Metonic) cycle and how to calculate Easter. He discussed general problems of time, measurement, computation, and astronomical phenomena and affirmed the practice of counting years from Christ's birth.

Prior to the eighth century, in spite of the decline in philosophical and scientific inquiry and the absence of any strong centralized government except in the Eastern empire, there was a varied economy and wide-ranging commerce. During the late seventh century, however, the explosive rise of Islam over the eastern Mediterranean and North Africa, and finally Spain, converted the Mediterranean into a Mohammedan lake and greatly restricted commerce. The cities declined in population, the merchant class dwindled into a few straggling peddlers, and European society moved steadily into feudalism. The eastern and western parts of the empire were separated, and although the Byzantine empire survived to the fifteenth century, it ruled a much reduced area. It was not until the Battle of Tours in 732 that the Islamic expansion was finally halted. (Others would argue that the Islamic expansion continued until the conquest of Sicily in 912.) A century earlier Alexandria had been conquered by the Arabs, and whatever part of the great library that had survived Christian fanaticism was destroyed. The famous story that the books were used by the Arabs to fuel the furnaces of the baths at Alexandria is probably apocryphal, although Caliph Omar is supposed to have asserted that either the books in the library were in agreement with the Koran, in which case they were superfluous, or they opposed the Koran and were heretical.

From 750–1000 the great patrons of science were the Caliphs of Baghdad; observatories were established at Baghdad and Damascus, and Ptolemy's *Almagest* was translated into Arabic in 820. By 1000 Spain had become an intellectual center of the Moslem world, and there the infusion of Greek thought into Christian Europe began. As contacts began to open up with the Spanish cities of Seville, Cordoba, and Toledo, knowledge of Islamic culture began to filter into Christian Europe. The library at Cordoba had 500,000 volumes when the Royal Library at Paris had 2,000. The introduction of trigonometry and "arabic" numerals (originally of Hindu origin), along with the sophisticated Ptolemaic models of the universe, astronomical instruments, and precise planetary records, led to a reawakening of interest in astronomy.

One obvious consequence was the infusion of a great many Arabic astronomical terms,[6] many of which survive today. Although Christian Europe would not challenge the sophistication of the medieval Arab astronomers until the fourteenth century, the beginnings came with the first contacts with Islamic scholars three centuries before. Thus the new science that crept into Western Christendom in the twelfth century was largely Arabic in origin, founded on the ancient Greeks, and learned directly from the Greek of the Byzantine empire and from Syriac, Persian, and Hindu sources, which themselves had Greek origins (figs. 7.1 and 7.2).

During the ninth and tenth centuries, when Islamic commerce and culture were at their zenith, European political and social structure was at its lowest ebb. In the cathedral and monastery schools Roman literature was becoming better known, but Greek as a language was still lost. By the beginning of the eleventh century, however, clear signs of recovery began to appear. The Norsemen and Magyars had been repelled, Christianized, and absorbed into European culture. Venice and Naples, and later Pisa and Genoa, reopened their Mediterranean trade routes and again became centers of commerce. Sicily, which had been a site of interchange of Greek, Arabic, and Latin ideas, was dominated by Islam until 1095, when it became a Norman settlement and provided favorable conditions for translations from Arabic to Latin. Toledo was recaptured by Christians in 1085.

From about the ninth century on, the sphericity of the earth and the geocentric system of planetary dynamics were reinstated in the minds of the clerical scholars, and after the mathematician Gerbert became Pope Sylvester II in 999, the spherical earth became once and for all an acceptable notion. Early in the twelfth century Platonistic scholars of the School of Chartres, the first of the great cathedral schools of the later Middle Ages, began to look for an explanation of the universe in natural causes; they embraced the homocentric sphere universe of Plato and Aristotle. Their cosmology was strongly influenced by the *Timaeus*, but Ptolemaic astronomy and Aristotelian physics were taught.

When, during the twelfth century, translations of Aristotle first began to reach France from Spain, the reaction of the church was initially hostile. Witness the following condemnation issued at Paris in 1210:

Neither the books of Aristotle on natural philosophy nor their commentaries are to be read at Paris in public or secret, and this we forbid under the penalty of excommunication.[7]

And yet it was not long before Aristotle came to be regarded as the most important of the philosophers and an ally of the scholastic theo-

Figure 7.1 Astronomers at Istanbul Observatory. Note the instruments in use, including an astrolabe and a quadrant, and the spherical globe of the earth. Sahinshshaname, Courtesy of Istanbul University Library.

logians. Fresh translations were made by the Dominican friar Albertus Magnus, one of the most original thinkers of the thirteenth century, and by his disciple Thomas Aquinas, who was to win even greater fame. Thomas Aquinas united Aristotle's cosmology and the doctrines of the church into one system of thought, which bearing the name

Figure 7.2 Thirteenth-century French illustration showing one astronomer making observations with an astrolabe, while another consults Arab astronomical tables. Bibliothèque Nationale, Paris.

Scholasticism dominated European minds for more than two centuries and still continues to influence Catholic theology. Thomas Aquinas accepted the rationality of science but denied that any necessity could be discovered in God. The works of Ptolemy were known to Aquinas, but perhaps Dante's *Divine Comedy,* with its multistoried universe, best expresses the general level of cosmography in the thirteenth century. In retrospect it can be said that the astronomical world view of the thirteenth century had finally risen to the level of Greek antiquity.[8]

The Foundations of Experimental Science

Aristotle can be seen as a sort of tragic hero striding through medieval science. From Grosseteste to Galileo he occupied the center of the stage, seducing men's minds by the magical promise of his concepts, exciting their passions and dividing their allegiances. In the end he forced them to turn against him as the real consequences of his undertaking gradually became clear; and yet, from the depths of his own system, he provided many of the weapons with which he was attacked.

A. C. Crombie[9]

During the twelfth century, translations of original Greek and Arabic philosophical and scientific works into Latin became widespread, so that by the end of the century most of Aristotle's works were available. These writings resurrected the Greek methodology of rational explanation and geometrical proof and stirred an interest in the natural world for other than didactic or symbolic purposes. Thus, at the School of Chartres, William of Conches and Thierry of Chartres made early attempts at rational explanations of Genesis. There was, of course, a tradition of Greek empiricism, exemplified by Aristotle's studies in biology, but no real experimental science had developed among either the Greeks or the Arabs.

Probably the first thoroughgoing unification of the twelfth-century traditions of technology[10] and logic, the first attempts to deal with the problems of induction and experimental verification, appeared in the work of Robert Grosseteste (c.1175–1253), a Franciscan, Chancellor of Oxford University and Bishop of Lincoln. Grosseteste adopted the two-edged sword of Aristotelian empiricism, with "inductive" generalization from experience to theory, followed by the deduction of consequences not contained in the original generalization, to be tested through experimental verification or falsification. He seems to have been the first to fully comprehend and apply this theory of experimental science. An essential ingredient in his description of phenomena was the use of mathematical abstraction. Grosseteste's views on scientific methodology had a particularly strong impact on the Oxford School.

Roger Bacon (c.1219–1292) absorbed the Grosseteste tradition while incorporating elements of his own, based in part on additional access to ancient sources. On the importance of experiment, he was very clear:

> This experimental science has three great prerogatives with respect to the other sciences. The first is that it investigates by experiment the noble conclusions of the sciences. For the other sciences know how to discover their principles by experiments, but their conclusions are reached by arguments based on the discovered principles. But if they must have particular and complete experience of their conclusions, then it is necessary that they have it by the aid of this noble science. It is true, indeed, that mathematics has universal experiences concerning its conclusions in figuring and numbering, which are applied likewise to all sciences and to this experimental science, because no science can be known without mathematics. But if we turn our attention to the experiences which are particular and complete and certified wholly in their own discipline, it is necessary to go by way of the considerations of this science which is called experimental.[11]

The scholastic rationalism of the Dominicans Aquinas and Magnus and the Franciscans Grosseteste and Bacon, combined with the practical traditions of the technics, led to a real scientific methodology characterized by a recognition of the inductive, experimental, and mathematical aspects of scientific inquiry.[12] Here was a systematic attempt to discover the nature of science; it led to a clear realization that results of science depend on sense experience, that is, are a posteriori in character.

The role of induction in science was explored by William of Occam (c.1285–1349),[13] who rejected causation and asserted that relations are concepts of the mind rather than objective principles of the cosmos, a view certainly incompatible with Aristotelianism. Occam's radical empiricism, in which it was asserted that what is not observed is not real, and his claim, with the Nominalists, that universals are mere names had an important influence on the growing spirit of empiricism in the fourteenth century. He is popularly known for his principle of economy of theory known as "Occam's Razor," one form of which he gave as "a plurality must not be asserted without necessity." Occam was strongly influenced by Grosseteste and Bacon, as was Magnus. Furthermore, the early founders of mechanics in the fourteenth century, at Oxford's Merton College and at the University of Paris, Thomas Bradwardine, Jean Buridan, and Nicole Oresme, were guided by the scientific methodology of Grosseteste, Bacon, Magnus, and Occam.

The Foundations of Mechanics in the Late Middle Ages

Mechanics in the twelfth to fourteenth centuries was dominated by the Aristotelian animistic tradition and its preoccupation with the distinction between natural and violent motion, intrinsic and extrinsic causes, and with questions of efficient and final causes. Yet in the midst of this era of Scholasticism, Aristotle began to come under criticism from philosophers of the Universities of Paris and Oxford. These criticisms were firmly grounded in the Aristotelian tradition, were framed in Aristotelian terms and consistent with his metaphysics, and yet the difficulties found to be contained in Aristotelian discussions of motion were profound, and the criticism not only weakened the position of Aristotelianism but also laid the foundation for acceptance of the Copernican theory.

The difficulty of understanding the persistence of motion in absence of continuing motive force troubled philosophers from before the time of Aristotle and was a principal preoccupation of the Dominican and Franciscan scholars of the fourteenth century. The modern view has been variously attributed to Galileo, Descartes, and Newton. Briefly, we now understand that, in the absence of an applied external force, uniform rectilinear motion will persist and that a force is required to *change* the state of motion of an object. The property of a body manifested in its resistance to acceleration or in its tendency to remain in uniform rectilinear motion is its *inertia*.

In the fourteenth century medieval mechanics began to take on a character of its own. Founded on the beginnings made by Archimedes, Philoponus, and others, it owed much to the earlier Scholastic philosophers of the twelfth and thirteenth centuries.

The two great centers of investigation into the laws of mechanics were the University of Paris and Oxford's Merton College. The Parisians are particularly celebrated for their attempt to grapple with the problem of inertia, which led to the concept of *impetus impressus,* an attribute of moving objects partaking variously of the ideas of inertia, momentum, and kinetic energy.[14] This idea may have been initially expressed by Philoponus, but it is first stated explicitly by the Parisian Jean Buridan. At about the same time Occam was rejecting the Aristotelian idea that a motive force was required to sustain motion, and yet he failed to introduce any intrinsic property to replace it, nor did he come close to the modern idea of inertia. Against the Aristotelian doctrine that the motive force resides in the air, which the process of propelling an object has disturbed, Occam noted that two archers could

arrange arrows to collide with each other from opposite directions, requiring that the air yield two different motions at once.

During the fourteenth century Buridan and his successors at the University of Paris developed a theory of dynamics in which impetus played
a major role. Buridan wrote:

> Therefore, it seems to me that it ought to be said that the motor in moving
> a moving body impresses [imprimit] in it a certain impetus or a certain motive
> force [vis motiva] of the moving body, [which impetus acts] in the direction
> toward which the mover was moving the moving body, either up or down, or
> laterally, or circularly. And by the amount the motor moves that moving body
> more swiftly, by the same amount it will impress in it a stronger impetus. . . .
> Finally that impetus is so diminished or corrupted that the gravity of the stone
> wins out over it and moves the stone down to its natural place. . . . When He
> created the World, God set each of the heavenly bodies in motion in the way
> that He had chosen—imparted to each of them an impetus which has kept it
> moving ever since. Thus God no longer has to move these bodies, except for a
> general influence similar to that by which He gives His consent to all things
> that occur. . . . The *impetus* that God imparted to the heavenly bodies is neither
> weakened nor destroyed by the passage of time. For in these heavenly bodies
> there are not tendencies towards other motions and because, moreover, there is
> no longer any resistance which could corrupt and repress this *impetus*. [15]

As a good Aristotelian, Buridan felt compelled to reconcile his idea
of *impetus* with the fact that violent motion had to have a violent or
extrinsic cause, whereas impetus would appear to be an intrinsic or internal property. Buridan argued the contrary, asserting that *impetus* is
efficiently violent. While foreshadowing the concept of inertia and containing the germ of the idea that motion continues unless interfered
with, impetus was still seen as a force or power moving a body, even
in the early writing of Galileo.

Buridan's Parisian student and contemporary Nicole Oresme (d.1382)
played a particularly crucial role in the history of astronomy and cosmology through his criticisms of Aristotle and Ptolemy. His arguments
provided a basis for accepting as plausible the idea that the earth moves,
and he succeeded in demolishing many of the physical arguments advanced by the Greek masters in favor of the geocentric universe. He
employed impetus in arguing that, if the earth moved, an arrow shot
vertically upward would, by virtue of the impetus impressed upon it
by the earth, return to the point of origin. Similarly the argument about
the arrow shot upward from a moving ship was answered by using
impetus theory, as were the supposed effects of a moving earth on the
air, clouds, and water. Oresme argued that the halting of the sun in

Joshua's day was an illusion and that the rotating earth must have been stopped instead. A rotating earth obviated the need for the outermost crystalline sphere. The appearances could yet be saved and the difficulty of the great velocities required for the sphere of the distant starts could be circumvented. The elimination of a crystalline sphere carrying the fixed stars, or at least the understanding of the plausibility of the idea, helped erode the idea that the universe is finite, limited by the outermost sphere (propelled by the primum mobile). It is not far to the "infinite spaces" of which Pascal was "terrified." In his Treatise on the Heavens and the World,[16] Oresme said,

Considering everything that has been said, one can then conclude that the earth is moved and the sky not, and there is no evidence to the contrary.

As early as the fourth century B.C. Heraclides of Pontus had argued that the earth's rotation could be used to explain the diurnal motion of the stars and planets and that the motion of the ninth or outermost sphere could thus be eliminated. This idea was known to scholars of eleventh-century Christendom as a result of the recovery of original Greek sources, even though it was not widely accepted. Heraclides and others had favored this hypothesis mainly on grounds of simplicity or from purely kinematical considerations. It is thus important that here for the first time, using the impetus theory, Oresme made the rotation of the earth plausible on physical grounds.

At about the same time somewhat similar and equally original studies of motion were being carried out at Oxford's Merton College, where John Brandwardine and others explored the laws of kinematics and dynamics.[17] They made the distinction between the causes of motion and the effects, and they explored the concepts of instantaneous velocity and of uniformly accelerated motion ("uniformly deformable motion"). Bradwardine and his colleagues studied the relationship between force or motive power and resistance, and in an attempt to overcome the difficulties in the Aristotelian idea that velocity is proportional to the ratio or force or motive power to resistance [$v \propto (F/R)$], devised a power-law formula that we would now represent as $v \propto \log (F/R)$. Perhaps the most famous result obtained by these Oxford scholars was the "Merton mean speed theorem," which established that in uniformly accelerated motion starting from rest the mean speed is one-half the final speed.

These studies at Paris and Oxford led, through the Italian and Parisian schools of the fifteenth and sixteenth centuries, to the mechanics of Galileo, while the criticisms of Aristotle by Buridan and Oresme foreshadowed Nicholas of Cusa and the heliocentrism of Copernicus.[18] When Copernicus was a student in Cracow, Buridan was required reading. Cusa (1401–1465), known primarily as a precursor of Copernicus, is credited with asserting that the universe has a center everywhere and a circumference nowhere. Although he stopped short of postulating an infinite universe, his proposition may be likened to the "cosmological principle" of modern cosmology.

Medieval Astronomy in Islam and Christendom

The Middle Ages represent a period of almost complete dormancy in the Latin West in both theoretical and observational astronomy. Even the twelfth-century Renaissance, containing the roots of modern science, did no more for a resurgence of interest in practical astronomy than lay the foundation. Yet this same period, from the eighth to the thirteenth centuries, saw the rise of Islamic astronomy to unparalleled heights. While it is true that Roger Bacon built an observatory at Oxford in the thirteenth century and that he just possibly may have built a rudimentary telescope and made observations with it,[19] it was not until the Renaissance of the fifteenth century that the empirical traditions of Scholasticism were coupled with astronomical theory derived from Islamic sources and the improved instruments of the period to yield useful observations.

In eighth-century Baghdad, Greek and Hindu astronomical writings, including the *Almagest,* were translated into Arabic. Observatories were established at Baghdad and Damascus, and ninth-century Baghdad became a center of scientific studies, the true heir to the glory of Alexandria. An important library of manuscripts was accumulated. Arabic numerals were introduced, and new astronomical tables were derived from those of Ptolemy with correction for precession. The Ptolemaic system was translated into a physical model of the universe by transforming the deferents and epicycles into spheres rolling inside one another, a concrete form much more satisfying to the realistic inclination of the Arab astronomers. In the tenth century one of the greatest of all Arab astronomers, Al-Battani (Albategnius), who died in 928, was studying precession, the progress of the sun's apogee,[20] and created the equivalent of tables of sines of angles. Measurements by Arab astronomers of the magnitudes of a great many stars are not without value even now, in that they permit studies of the changing brightness over the past 1,000 years. By the late tenth century, Cordoba had taken over from Baghdad as the intellectual center of the Moslem world, and the earliest translations from Arabic into Latin date from this period. In the thirteenth century the power of Islam in Spain ended, and under the Christian King Alfonso X a set of astronomical tables, the Alphonsine Tables, were derived, which served until the time of Kepler. During the same period an elaborate observatory was established at Maragha in Persia with many important astronomical instruments and a library with 400,000 manuscripts.

Translations were carried out at Toledo, Cordoba, Seville, and other centers, with activity reaching its peak in the twelfth century. At Toledo Gerard of Cremona (1114–1187) translated Euclid, Apollonius, Ptolemy, Galen, and other Greek authors into Latin, as well as the work of Arab astronomers and commentators such as Ibn Sina ("Avicenna," d.1037) and Ibn Rashd ("Averroes," d.1199). The *Almagest* was translated into Latin by Gerard in 1175,[21] and as early as the tenth century treatises on the use of the astrolabe had been rendered into Latin. Thus by the end of the twelfth century, much of the heritage of Greek and Islamic astronomy was accessible to the scholars of western Europe.[22]

Astronomical Instruments of Antiquity

All of the astronomical instruments of antiquity were available to the Arab astronomer, including the gnomon or sundial in its various guises, armilla of several forms, the triquetrum (fig. 7.3), or rule of Ptolemy, and the astrolabe. Most of these instruments were known to Ptolemy and several date back to the third century B.C. or earlier. The only instrument possessed by the Greeks of the sixth century B.C. was the vertical gnomon. As we have seen, the gnomon was probably known to all civilizations of antiquity. The Chinese are said to have used a gnomon to determine the obliquity of the ecliptic (due to the inclination of the earth's axis) in 1100 B.C., and an early reference to a sundial may be found in Isaiah (38:8). With the gnomon it was possible to tell time; that is, it functioned as a primitive sundial or shadow clock and was used to find the meridian and to determine the date of the equinoxes and solstices.

The gnomon remained virtually the only astronomical instrument until the late fourth century B.C., when a primitive diopter, an instrument similar to the modern transit, seems to have come into use. At about the same time, the spherical sundial was devised by Eudoxus or Apollonius. Probably the diopter was first used for measuring angles by Hipparchus, in the second century B.C. The plane astrolabe (fig. 7.4) is often attributed to Hipparchus, and he seems to have discovered the

Figure 7.3A Triquetrum (three-staff) or Ptolemy's rule. With such an instrument Copernicus himself made observations which contributed to the demise of the geocentric universe. The arm A-A' was graduated and the observer's sights were located on the arm B-B'. R. T. Gunther, *Early Science in Oxford*, vol. 2 (Oxford, 1928).

stereographic projection technique on which it is based. Nonetheless it is not mentioned by Ptolemy in the *Almagest*, and we can be sure only that the astrolabe was used by Arab astronomers and navigators.

After the decline of Alexandria, new developments in observational instruments were made by the Arabs from the eighth century onward. In particular, the *meridian circle* was introduced, and the *plane astrolabe* was developed into an instrument of great utility and often of great beauty. In the twelfth century the torquetum (fig. 7.5) an equatorial sighting device with motion about a polar axis in the equatorial plane, was introduced, and in the fourteenth and fifteenth centuries the cross-staff was used by Regiomontanus and others for measuring angles. Further development of astronomical instruments consisted principally of refinements in graduation and increase in size, and gradually serious astronomical observations came to be made primarily with large quadrants, sextants, and meridian circles.

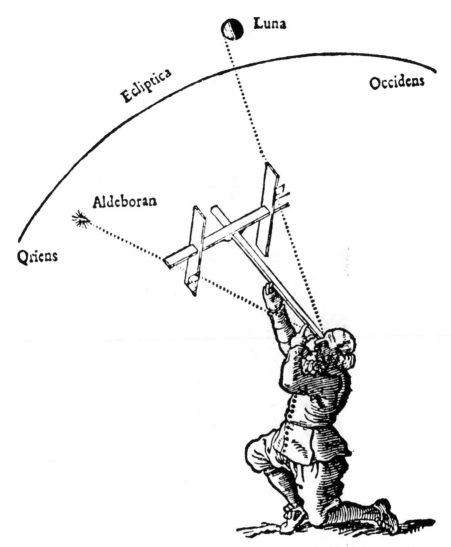

Figure 7.3B Astronomer making observations with a cross-staff or Jacob's Staff. Bayerische Staatsbibliothek, Munich.

The accuracy of pretelescopic astronomical observations depended strongly on the dimensions of the instrument employed in very much the same way that the resolution of an optical or radio telescope is a function of its aperture. It is also true, of course, that the gradual evolution of a tradition of careful observation was crucial to any progress in astronomical measurement and that large instruments, in themselves,

Figure 7.4 Late-fourteenth-century Arab astrolabe. The face of the astrolabe (rete) is a projection of the sky onto a plane. The names and positions of about 28 stars are given (each point of the cutout rete points to a star). The back of the astrolabe was a sighting device consisting of an alidade and altitude graduations. Gunther, *Early Science in Oxford,* vol. 2.

are no guarantee of precision. An understanding of the role of the dimensions of an instrument is demonstrated in the development of ever larger sundials and quadrants during the late medieval period. While the instruments available to Ptolemy and his contemporaries had dimensions on the order of 10 to 100 centimeters, by the fifteenth and sixteenth centuries quadrant circles with radii of several meters were not uncommon. If we consider the triquetrum, for example, we find that an error of a few millimeters in the position of the eye on the sighting rod would introduce an error in the measured angle of about

Figure 7.5A A torquetum, an equatorial sighting device, with the lower graduated circle being in the plane of the celestial equator. Purchased by Nicholas of Cusa in 1444. Cusanus-stift, Bernkastel-Kuis.

1/6 degree or 10′ of arc. Since the unaided eye can at best resolve sources that are 2′ apart, the accuracy achievable with a triquetrum of average size was poorer by five times than the limit imposed by the human eye. By constructing an instrument with a characteristic dimension of several meters, that is, more than ten feet, the maximum precision could be achieved. A further increase in size will bring no additional accuracy, except as it permits easier, and thus perhaps more reliable, observations. The huge gnomons and quadrants of the fabled Persian and Indian observatories of the fifteenth century (fig. 7.6) were much larger than necessary if observational accuracy were the principal criterion. But in fact, these instruments were as much works of architecture as of

Figure 7.5B An armilla or armillary "sphere" built for practical observation by Arab astronomers. Istanbul University Library.

astronomy. The largest known quadrant was one of 180 feet (54 meters) at Samarkand in Central Asia, in what was formerly eastern Persia.

The fine instruments built by Tycho Brahe (chapter 9) in the late sixteenth century, owing to their substantial size and finely divided circles, permitted regular observations whose accuracy was limited only by the resolution of the eye. They represent the culmination of the long history of pretelescopic astronomy.

Slowly interest in practical astronomy was awakened through the infusion of the ideas of the Greek and Arab astronomers and philosophers and as a result of the recovery of the astronomical instruments of antiquity, and by the end of the fourteenth century astronomical observations worthy of preservation were being made. Thus, although they contributed little to the development of theory, such fifteenth-century observational astronomers as George Purbach, his pupil Johann Muller

Figure 7.6 An elaborate solar observing device at the eighteenth-century New Delhi observatory built by Jai Singh. J. Cortazar, *Prosa del Observatoria* (Barcelona: Editorial Lumen, 1972).

or Johannes de Monte Regio (Regiomontanus),[23] and Paolo Toscanelli revived the largely forgotten concern with precise astronomical measurement and calculation, improved the tables of Alphonse X, and laid the groundwork for the achievements of Copernicus and Tycho. Toscanelli, who stands as perhaps the chief observational astronomer of the fifteenth century, was geographical adviser to Columbus and was a contemporary of Nicholas of Cusa at Padua, a center that was to have a profound influence on Galileo. Toscanelli is particularly noted for his precise observations of comets from 1433 to 1472 and for his studies of precession (although his observations were not discovered until 1864).

In science as in the larger history of man, the Middle Ages were a period of transition between the world of antiquity, typified by the achievements in Greek rational inquiry and dominated by Platonic and Aristotelian philosophy, and the empirico-deductive methods of modern science. It was a period not without real progress in science, yet much of it was spent in recovering the Hellenic traditions lost during the disintegration of the Roman Empire. Finally in the late Middle Ages, with the rise of Scholasticism, man's right to confront the universe rationally was reasserted, if timidly. Spurred by the infusion of Greek philosophy, mathematics, and astronomy, not only did the modern scientific method begin to take shape, but also substantive understanding of the physical world, particularly in biology and mechanics, was achieved.[24] The Middle Ages merged into the humanism of the Renaissance, exhibiting humanity's renewed confidence in its ability to unravel the secrets of creation. Thus in the revolution set in motion in the mid-sixteenth century by Copernicus and crowned by Newton and the Enlightenment in seventeenth- and eighteenth-century Europe, we see the final establishment of a world view whose roots lay in late medieval times.

CHAPTER 8

The Copernican Revolution, 1

Finally we shall place the Sun himself at the center of the universe. All this is suggested by the systematic procession of events and the harmony of the whole Universe, if we only face the facts, as they say, with both eyes open.[1]

Nicholas Copernicus

The resumption of astronomical research and of speculation in dynamics had been long in coming, but during the sixteenth and seventeenth centuries innovation was almost as rapid as one could have thought possible. This time of change is epitomized by the work of Nicholas Copernicus, who first used the heliocentric system in astronomy, and that of just four other men:[2] Tycho Brahe, the greatest of pretelescopic astronomical observers; Johannes Kepler, who discovered the true motions of the planets; Galileo Galilei, who brought the telescope to astronomy and systematic inquiry to dynamics; and Isaac Newton, who found a universal solution to the problem of motion. Because the cosmology that resulted is close to that of the present day, these men are often treated as moderns, rather than as the transitional figures they were. A helpful simile is used by Kuhn,[3] who says that an innovator, like the bend in a road, belongs to both the before and the after, to the old tradition and the new. Here in any event was something more complex than the traditional picture of uninterrupted scientific progress. After Copernicus, there would be no doubt that something needed to be done about the problem of the planets, but the attacks on the problem were as diverse as the nationalities and temperaments of the men themselves. Even the three contemporaries, Tycho, Kepler, and Galileo, found little enough to agree upon. In tracing the Copernican system to its issue as the "Newtonian world view" we shall therefore be learning not only the natural philosophy of the period but also something about how science is done.

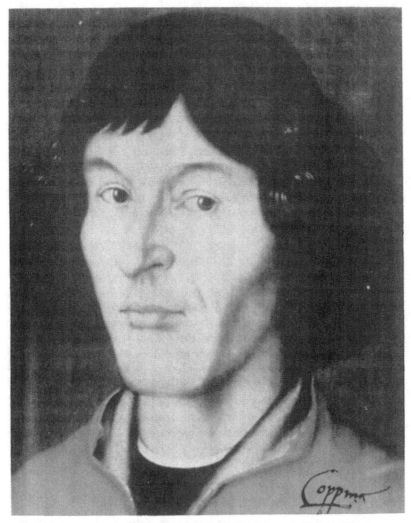

Figure 8.1 Nicolas Copernicus (1473–1543). Cracow University Library.

. The Italian renaissance was nearly a hundred years old when Copernicus was born in 1473 in Poland. The rediscovered Latin and Greek
classics had deeply influenced the church and the universities. The "new
learning" had reached the North in a less pure form, but its humanistic
effect was all the more influential because the level of Northern culture
had been so low. Copernicus, nephew and ward of the Bishop of Erm-
land, was the beneficiary of both the Northern and Southern humanism. His future assured by his uncle's position—the bishop was in effect

prince of the province—Nicholas spent fourteen years at various universities in acquiring a properly rounded education. He studied for four years at Cracow. Then, after being appointed at 23 a canon of the cathedral at Frauenberg, he spent ten years at Italian universities, Bologna and Padua and briefly Ferrara. By the time he returned to Poland with his doctorate in canon law, he had studied not only philosophy and law, but also Greek, medicine, and mathematics, and his interest in astronomy had been sharpened to expertise. His duties at Frauenberg certainly did not require such erudition. This is indicated by Copernicus' absence for the first ten years of his appointment and by his return, not to Frauenberg, but to the bishop's residence, Heilsberg Castle, where for six years he was in attendance to his uncle as legal advisor, secretary, and physician.

By 1513, when Copernicus finally took up residence at Frauenberg, he had already formulated the Copernican system. It was characteristic of Copernicus, a somber and secretive person, that he published nothing at this time. He did circulate his ideas among his friends in the form of a brief manuscript, the *Commentariolus*. The few copies of this document achieved a wide circulation, and Copernicus became known as a theoretical astronomer of the first rank. In the *Commentariolus* Copernicus wrote, "I have thought it well, for the sake of brevity, to omit from this sketch mathematical demonstrations, reserving these for my larger work." The larger work referred to, *de Revolutionibus Orbium Coelestium,* finally reached print while Copernicus lay on his deathbed. Earlier, in 1539, Copernicus had allowed a young mathematician, Rheticus, to publish an account of his work, called the *Narratio Prima,* but permitted no mention of his name. This reticence, besides being a natural part of Copernicus' personality, is understandable on religious and political grounds. Though Copernicus never took holy orders, he was sensitive to his position within the church and scrupulously avoided controversy. As we shall see when we examine the structure of *de Revolutionibus,* Copernicus had no intention of overthrowing or threatening theological doctrine; in fact he dedicated the book to Pope Paul III. Whereas it is perhaps too much to expect of Copernicus that he should have thought the displacement of the earth from the center of creation a dangerous idea, others of his time were more perceptive. Martin Luther, for example, condemned Copernicus during the latter's lifetime, and the Copernican system quickly became anathema to the Protestant Reformation. Luther said, in his "Table Talks,"

People gave ear to an upstart astrologer who wanted to prove that the earth was moving and revolving, rather than the heaven or the firmament, sun, and moon; just as if someone in a moving carriage or on a sailing ship believed that he was motionless and in rest, but that the earth and trees were moving. But such are the times we live in; he who wants to be clever must invent something all his own and what he makes up he naturally thinks is the best thing ever! This fool wants to turn the whole art of astronomy upside down! But as the Holy Scripture testifies Joshua bade the sun to stand still, not the earth![4]

The Catholic Church for various reasons was late in rising to condemn Copernicus, and it was not until 1616, during the Counter-Reformation, that his works were proscribed. General opposition might have developed much earlier had not *de Revolutionibus* been so technically difficult. By the time lay and clerical authorities began to raise substantial opposition, the book had become an essential work for enlightened European astronomers.

The Development of the System

Copernicus objected to the Ptolemaic system because it was arbitrary and unsymmetric. As he wrote, "a system of this sort seemed neither sufficiently absolute nor sufficiently pleasing to the mind."[5] For Copernicus the only "sufficiently pleasing" motions were compounded from uniform circular motion. The Ptolemaic system used epicycles and eccentrics, which produced acceptably regular motion by this definition. But nearly all versions of the system—Ptolemy's own as well as later ones[6]—used the equant, which gave a variable angular speed about the center of the circle along which the epicycle moved. This lack of symmetry led Copernicus to examine "whether there could perhaps be found a more reasonable arrangement of circles, from which every apparent inequality would be derived and in which everything would move uniformly about its proper center, as the rule of absolute motion requires." The very arbitrariness of the method of compounded circular motion must have encouraged Copernicus to look for a new system. Ptolemy had claimed no uniqueness for his calculations, and there were many versions of the system. Of this Copernicus wrote in the preface to *de Revolutionibus:*

nothing except my knowledge that mathematicians have not agreed with one another in their researches moved me to think out a different scheme of drawing up the movements of the spheres of the world.[7]

There is of course no way to know how Copernicus came to believe that the earth moves. The statements quoted above indicate that he probably tried the simplest heliocentric system—stationary sun, circular orbits, rotating earth—as one of several possible modifications to the Ptolemaic system (fig. 8.2). What we do know is that this revolutionary idea was not demanded by accumulating evidence of the inadequacy of the Ptolemaic system, as is sometimes argued ("Ptolemy's planets, playing fast and loose, foretell the wisdom of Copernicus").[8] There simply was no external reason, in the form of new observations, new data, that required a new theoretical approach. It is undoubtedly important that Copernicus knew of Aristarchus'[9] ancient attempt, but perhaps equally important were the Neo-Pythagorean inclinations, which were widespread during the Renaissance and which bordered on "sun

Figure 8.2   The simplest possible heliocentric system, with the planets moving in circular orbits around the sun, explains in an approximate way all the features of planetary motions. The symmetry of this idealized system was especially appealing to advocates of circular motion, such as Copernicus and Galileo. On some occasions they must have been thinking of this arrangement of circles while they were arguing for the full Copernican system. From the manuscript of Copernicus de Revolutionibus. By permission of Houghton Library, Harvard University. Photo by Owen Gingerich.

worship," with their echoes of Philolaus' cosmology of an earth moving about a central fire. Copernicus indeed referred to the Pythagorean tradition of passing truth by word of mouth as justification of his own reluctance to publish. In any event he must have been struck by the simplifications that follow from the simple, symmetric system of circles concentric about the sun:

1. The retrograde motions of the planets are seen to be entirely due to the earth's motion. When Mars, Jupiter, and Saturn are overtaken by the earth, they reverse their directions of motion. For Mercury and Venus the retrogradations occur when they catch the earth.

2. The association of Mercury and Venus with the sun is no longer accidental, as it is in the Ptolemaic system. Rather, they stay near the sun in the sky because they are near the sun in space. The one-year geocentric (or "zodiacal") period for these inner planets is explained at the same time; any planet whose orbit is inside the earth's has the sun as its average position.

3. The rotating earth allows the stars to be stationary. Many philosophers had objected to the immense velocities required for the sphere of stars to complete one rotation a day. Copernicus was bothered less by this than by the problem that if the earth is stationary, precession of the equinoxes requires that the heavens themselves are not perfectly regular in their motions. For Copernicus the problem seemed even more acute, since he believed, quite incorrectly, that the rate of precession had varied since ancient times. In any event, precession could now be explained in terms of the motion of the earth.

Of course, the planetary orbits are *not* circular, so there remained the problem of modifying the simple heliocentric system to obtain detailed agreement with the observed positions of the sun, moon, and the five planets. This could only be accomplished, Copernicus thought, by using epicycles and eccentrics (but not the hateful equant). It was for this purpose that he wrote *de Revolutionibus*. He devoted four of the six books of that work to calculations. And in the calculations he became a Ptolemaic again, at least in method, bogged down in detail and forgetting his objections to the awkwardness of the earlier system.

Displacing the earth and halting the heavens did grave violence to Aristotelian cosmology. Nothing could save it, but Copernicus tried lamely to retain the Aristotelian language. The philosophers had ultimately rejected the rotation of the earth, because as a violent motion it would have caused the earth to "become dispersed" and "pass beyond the heavens." Since this has not happened, Copernicus argued, and since rotation does occur, rotation must be a *natural* motion, not a violent one. After all, he added, for a sphere "the movement which accords naturally with its form" is rotation. In this bit of sophistry Copernicus used "natural" in two different senses, and unconsciously exposed the shallow roots of the Aristotelian physics.

Having thus established, to his own satisfaction at least, that the earth can rotate, Copernicus asked himself whether the earth can leave the center of the universe. In Aristotle's system the element earth had gathered at the center of the universe, which also served as the center for planetary motions. Copernicus suggested that, since the earth is not at the center of motion, perhaps gravitation is associated with the spherical shape rather than with the center of the universe:

For the apparent irregular movement of the planets and their variable distances from the Earth—which cannot be understood as occurring in circles homocentric with the Earth—make it clear that the Earth is not the center of their circular movements. Therefore, since there are many centers, it is not foolhardy to doubt whether the center of gravity of the Earth rather than some other is the center of the world [universe]. I myself think that gravity or heaviness is nothing except a certain natural appetancy implanted in the parts by the divine providence of the universal Artisan, in order that they should unite with one another in their oneness and wholeness and come together in the form of a globe. It is believable that this effect [i.e., gravity] is present in the sun, moon and other bright planets and that through its efficacy they remain in the spherical figure in which they are visible, though they nevertheless accomplish their circular movements in many different ways. Therefore if the Earth too possesses movements different from the one around its center, then they will necessarily be movements which similarly appear on the outside in the many bodies; and we find the yearly revolution among these movements.[10]

Copernicus was arguing that the earth is like the planets in shape and in motions; he said in several places that it can be regarded as one of the wandering stars. But he stopped short of examining the broader implications of making the earth one of many similar bodies.

In the *Commentariolus* Copernicus baldly stated the central role of the sun:

All the spheres revolve about the sun as their mid-point, and therefore the sun is the center of the universe.[11]

And in the opening Book of *de Revolutionibus* Copernicus celebrates the dominant position of the sun.

In the center of all rests the sun. For who would place this lamp of a very beautiful temple in another or better place than this wherefrom it can illuminate everything at the same time? As a matter of fact, not unhappily do some call it a lantern; others, the mind and still others, the pilot of the world. Trismegistus calls it a "visible god;" Sophocles' Electra, "that which gazes upon all things." And so the sun as if resting on a kingly throne, governs the family of stars which wheel around.[12]

Not much physics here, but this kind of Neopythagorean language was common in the late Renaissance. Some writers have called Copernicus and his followers sun-worshippers because of their allegorical references to the sun. In the calculations of *de Revolutionibus,* however, the sun is finally deprived of much of its centrality. It remains as the only stationary object below the stars, but the orbits of the planets are referred to the moving center of the earth's orbit. It is as if the vision of symmetry is obscured by the details. It remained for Kepler, for whom the Pythagorean dream was an obsession, to show how the motions of the planets could be referred to the sun alone, without any preferred role for the earth.

Among the traditional reasons for rejecting the wandering earth was the belief that the universe is simply not that big. The sphere of stars was thought to begin near the orbit of Saturn; if irregularities in the positions of the planets result from the earth's motion, then the stars should also reveal the earth's motion by moving relative to each other. (fig. 8.3). Since no such parallax was observed, Copernicus asserted that the stars must be vastly more distant than the planets. Ptolemy had discussed the fact that the horizon divides the celestial sphere equally— an effect that implies that the heavens are much larger than the earth— and had argued that the earth must therefore be at the center of the universe. Copernicus used the same fact as an indication that the whole of the earth's orbit is small compared to the distance to the stars.

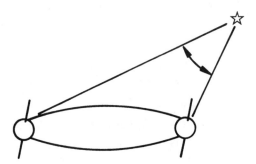

Figure 8.3 Stellar parallax illustrated schematically. If a star were near the earth, its apparent position in the sky would change as the earth moved in its orbit. With modern telescopes a small amount of parallax can be resolved for the nearer stars.

For that demonstration [bisection of the celestial sphere by the horizon] proves nothing except that the heavens are of an indefinite magnitude with respect to the Earth. But it is not at all clear how far this immensity stretches out. On the contrary, since the minimal and indivisible corpuscles, which are called atoms, are not perceptible to sense, they do not when taken in two's or in some small number, constitute a visible body; but they can be taken in such a large quantity that there will at last be enough to form a visible magnitude. So it is as regards the place of the earth; for although it is not at the center of the world, nevertheless the distance is as nothing, particularly in comparison with the sphere of the fixed stars.[13]

Another problem with removing the stars to great distances is that to the eye the brightest stars look like they are about the same angular size as the planets, some two minutes of arc (1/30 of a degree). This apparent size is due to the limited resolution of the eye; the true angular diameters of the stars are very much smaller. But no one knew this until the telescope came along in the seventeenth century. If the stars were as large as their apparent size, they would be immensely larger than the sun. This had been one reason for the rejection by the Greeks of the heliocentric world suggested by Aristarchus: Toulmin and Goodfield have commented concerning this point.

The general public today is so innured to astronomical statistics that people will say unthinkingly, "The stars are an incredible distance away"—and leave matters at that. Greek astronomers were more particular about their arguments, and declined to believe the unbelievable.[14]

Copernicus offered no fully convincing arguments for the validity of his system, and there were few men who were willing to "believe the unbelievable." The survival of the Copernican system during the second half of the sixteenth century depended on its use by astronomers who found it more convenient than the Ptolemaic system but who were interested only in saving the appearances and not in the question of the earth's true motion.

On the Revolutions of the Heavenly Spheres

Copernicus modeled *de Revolutionibus Orbium Coelestium* after Ptolemy's *Almagest*.[15] The work was intended not only to persuade the reader of the validity of the heliocentric system but also to provide the full mathematical demonstration promised in the *Commentolarius*. Moreover, the results of calculations of solar, lunar, and planetary positions were displayed in tables that were immediately useful to the practicing astronomer, or astrologer. The agreement with observation was on the whole no better than Copernicus would have obtained using the Ptolemaic system, but the calculations were fresh and hence free of the cumulative errors that grow out of imperfect knowledge of planetary periods and other constants. Copernicus accepted uncritically the measurements recorded in the Greek and Arabic tables. He wrote, regarding discrepancies among earlier authorities,

it is believed by many that some error had crept into their observations. But each mathematician is alike in his care and industry, so that it is doubtful which one we should follow in preference to another.[16]

This attitude led Copernicus to introduce complicated explanations for effects that were not real.

The organization of de Revolutionibus is indicated in table 8.1. As we have seen earlier, Copernicus was eager to have the reader accept the reasonableness of the heliocentric hypothesis. The remarks preceding the preface deny, however, that the Copernican system is better than the Ptolemaic. The language used in this beginning section is so different from that in the body of the work that it is not surprising to learn that the writer was not Copernicus, but Andrew Osiander, who supervised the first printing of de Revolutionibus. Osiander wrote,

The author of this work has done nothing which merits blame. For it is the job of the astronomer to use painstaking and skilled observation in gathering together the history of the celestial movements, and then—since he cannot by any line of reasoning reach the true causes of these movements—to think up whatever causes or hypotheses he pleases such that, by the assumption of these causes, those same movements can be calculated from the principles of geometry for the past and for the future too. This artist is markedly outstanding in both these respects: for it is not necessary that these hypotheses be true, or even probable. . . . Maybe the philosopher demands probability instead; but neither of them will grasp anything certain or hand it on, unless it has been divinely revealed to him.[17]

Perhaps some clerical authorities thought Osiander's words were Copernicus', and were pacified. Copernicus himself nowhere compromised his belief in the moving earth, not even in the dedication to the Pope, where he asserted that mathematicians would be forced to accept his system,

if—as philosophy demands in the first place—they are willing to give not superficial but profound thought and effort to what I bring forward.[18]

Table 8.1. Abridged Table of Contents

On The Revolutions of the Heavenly Spheres

He appealed to the Pope as the most eminent lover of letters to "provide a guard against the bites of slanderers" who "by shamelessly dis-
torting the sense of some passage in Holy writ" would criticize the
system. But he professed not to be worried about the judgments of the
unlearned, since "mathematics is written for mathematicians." There is
something of the Pythagorean tradition in this last, but the confidence
that the church would recognize the truth of scientific discoveries is
evidence that Copernicus was no iconoclast. Almost a century later
Galileo made a similar appeal to the church as repository of truth; by
then the response was immediate and hostile.

Most of the body of *de Revolutionibus* is far too technical to be described here. We shall examine only a few of the details of the planetary
motions, enough at least to demonstrate the complexities forced on
Copernicus. The earth's motion was treated first. The path of the center
of the earth is along an eccentric circle, as shown in figure 8.4. This is
equivalent to the Ptolemic scheme (except, of course, that for Ptolemy
the *sun* moved on an eccentric circle) with an added feature, the small
eccentric circle *CFE,* which accounted for a small change in the orientation of the earth's orbit. Whenever disagreement among several measurements forced an added element of motion—as in the case of the
shifting orbit—Copernicus always assumed the change was periodic and

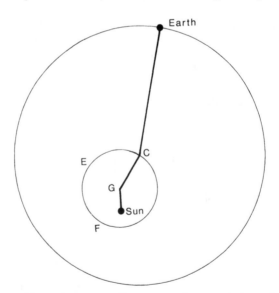

Figure 8.4A The earth's revolution according to one of two equivalent schemes proposed by Copernicus. The center *C* of the earth's circular motion travels around the circle
CFE in 3400 years.

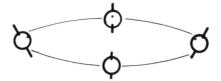

Figure 8.4B Copernicus accepted the Ptolemaic idea that revolution occurs as if the re-volving body were connected to its center of revolution by an imaginary spoke, so that the same "face" is always turned toward the center. For the moon this actually happens. For the earth, this "rolling" motion would cause the polar axis to sweep out a cone annually. To cancel the conical motion, Copernicus added a conical motion in the oppo-site sense.

would finally repeat itself, so that it could be explained by compounded circular motion. The alternative was to admit that the heavenly motions were truly irregular and therefore inexplicable, or so Copernicus thought.

The model for the orientation of a planet moving on an epicycle was the moon, which always turns the same face to the earth and so moves as if embedded in a wheel (or a sphere). The lunar motion was called motion without rotation, although in fact the moon makes one rotation for each revolution. This misconception about rotation made the Co-ernican description of the earth's rotation rather cumbersome. The earth's motion on its eccentric would cause the earth's polar axis to trace out a cone (fig. 8.4B). This is not observed; the earth's axis is almost constant in direction. Copernicus therefore introduced an additional motion to counteract the expecting "rolling." If the two motions exactly canceled, the earth's axis would always point in the same direction. The phenom-enon of precession of the equinoxes occurs because the direction of the earth's axis shifts slowly. Copernicus explained precession by making the period of the antirolling motion slightly greater than the period of the earth around the sun.

The period of precession is 26,000 years (chapter 2). Various esti-mates of this interval had been obtained during the preceding 1,600 years. In a misguided attempt to reconcile all of these values Copernicus added two new motions for the poles of the earth. The geometrical device he introduced to account for variation in the rate of precession is called "libration," and it is curiously uncircular.[19] Copernicus was dealing with small corrections to the length of year, about 1 part in 100,000, which would correspond to about 1 day in 300 years. In the view of Copernicus these uncertainties were excessive, and meaningful calendar reform would require further study.

The Copernican lunar theory used an epicycle on an epicycle, centered, of course, on the earth. The only significant change from the Ptolemaic moon was an improved treatment for the range of distances to the moon, which is related in the Copernican system to the sizes of the orbits of the planets. The mean earth-moon distance (in units of the radius of the earth) is used to determine the mean earth-sun distance, and from this the distances of the other planets are found. Copernicus obtained accurate values for all these distances.

Copernicus assumed, as Ptolemy had, that the motions of the planets in longitude could be treated separately from their motions in latitude. Since the basic features of the planetary motions—their mean periods and their retrogressions—were explained by the earth's motion, the subtle features Copernicus described in Book V (Longitudes) and Book VI (Latitudes) were all in the nature of corrections. The basic scheme for the planetary longitudes was as shown in figure 8.4; the planet moved on a small epicycle whose deferent was centered on an eccentric. The eccentric was, however, referred to the center C of the earth's circle (see fig. 8.5) rather than directly to the sun. This could be interpreted as giving the earth a preferred role, although Copernicus could have completed the sun-centered description for each planet by adding the circle CFE and the eccentric GS. This system was satisfactory for the superior planets, Saturn, Jupiter, and Mars; for Venus and Mercury other corrections were necessary. Copernicus added a moving eccentric to the description for Venus. Mercury was particularly intractable; for it the full system—a librating epicycle on an eccentric circle on an eccentric referred to the earth's moving eccentric—is furthest of all from the "much simpler constructions" promised earlier.

The motions in latitude were treated in full and unsatisfactory detail in the last book of *de Revolutionibus*. All the planets move in planes near the ecliptic, so the deviations from the ecliptic are related in a simple way to the longitudinal positions of the planets. Copernicus had no idea what caused the planets to move as they did, and his values for planetary latitudes contained many errors. He was led far afield, proposing complex combinations of epicycles and librations on the basis of only a handful of measurements. As before, the motions were referred to the center of the earth's orbit, and the periods of the latitudinal motions were simple multiples of the earth year. If referring the planetary epicycles to the earth's orbit was an excusable shortcut to referring each to the sun, it was a form of geocentrism to make their periods commensurate with the year. Or perhaps the related values were forced on Copernicus by the limited data available to him, since he had too few observations to determine the parameters of the complex orbits he assigned, if all the radii and periods had been assigned independently.

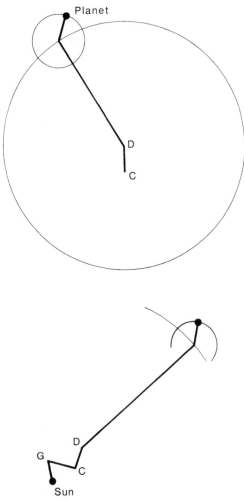

Figure 8.5 Saturn and Jupiter, according to Copernicus, moved on a small epicycle whose deferent was eccentric by the distance *CD* from the center of the earth's orbit. Referred to the sun, the scheme was in effect that of the lower diagram. (The notation is that of the previous figure.)

This catalogue of the eccentrics, epicycles, and librations of the full Copernican system gives undue emphasis to minor corrections and to superfluous features that could be ignored by later astronomers. For the typical astronomer the tables provided in *de Revolutionibus* were satisfactory. The astrologers could skip the most complicated sections, since in astrology latitudes are ignored altogether. The question of the correctness of the Copernican system did not concern the sixteenth-century practitioner, and in any event little progress could be made without more accurate and more numerous observations. Tycho Brahe, who first realized the importance of mapping the full orbits of the planets, made the measurements that were capable of allowing an unambiguous choice between the Ptolemaic and Copernican systems, and Johannes Kepler showed how the choice was to be made.

CHAPTER 9

The Copernican Revolution, 2

Copernicus built his elaborate heliocentric system for the planets on a handful of his own measurements of planetary positions (fewer than fifty observations in his entire life) together with whatever published measurements came to hand. There were two failures of sorts involved in this. First, there was Copernicus' uncritical, unwarranted confidence in the accuracy of all measurements ("Each mathematician is alike in his care and industry"[1]). Secondly, Copernicus and all who preceded him were victims of a particular trap arising from the fallacy of compounded circular motion. The trap was this: since one and only one circle can be drawn through three points, three observations suffice to determine a circular orbit, six at most are required for an epicycle and a deferent, and so forth. Even the most complex of "circular" constructions for a planet can be derived from a dozen or so measurements, provided the measurements are made with sufficient accuracy—the first point again. Copernicus can be pardoned for not grasping the concept of uncertainty in measurement. The problem is a subtle one, and the mathematical tools ("statistics") for dealing with experimental errors would not be developed until the nineteenth century. Even the genius Isaac Newton, working 150 years after Copernicus, floundered when it came to dealing with very small differences between his measured results and his predictions.

Of course it simply isn't true that all measurements are equally accurate. Copernicus knew that, at least at some level. Perhaps he thought that measuring is like counting. Anyone who is careful can correctly find the number of coins in a jar, and, because the answer is unique, it is not influenced by the method of counting. Measurement of an angle, or other continuously variable quantities, is not like counting.

The method does matter, and what is accurate enough for one purpose may be useless for another. What is obtained, in fact, is a range of values within which the "true" angle (or other variable) is very likely to be. Modern methods can tell the experimenter what this range, the "uncertainty," is in a given case and can suggest ways to improve the measurement by reducing the uncertainty. But what could be done in the sixteenth century?

Tycho

Tycho Brahe (fig. 9.1), working from the 1570s to the end of the century, did the best that could have been done in the pretelescopic era. By systematically observing with large, finely crafted instruments he obtained observations limited only by the resolution of the eye. Tycho's data for planetary positions were good enough, barely, for Johannes Kepler to derive the true figure of motion for the planets, the ellipse. Tycho's dedication to observation, amounting almost to fanaticism, led him out of the "handful of measurements" trap. Tycho made repeated observations of the planets; he and his assistants recorded the positions of the planets on virtually every clear night over a period of more than ten years. It is not easy to find a sufficient motive for an effort on such a heroic scale. An incident that certainly influenced Tycho occurred when he was just beginning to learn astronomy as a youth of 17. He found that the date for a conjunction of Jupiter and Saturn was given wrongly by both the Alphosine Tables (Ptolemaic) and the Prussian Tables (Copernican). Tycho was no Copernican, and he came to disbelieve the Ptolemaic system as well. Later he proposed his own planetary model,[2] the Tychonic system (fig. 9.2), which assumed that the sun and the moon revolve about the earth while all the planets revolve about the sun.

Tycho had no peers. No theorist guided him in his efforts, no spirit of competition spurred him. Yet, without Tycho's measurements Kepler could not have found his planetary laws, and the course of seventeenth-century science might have been very different. From the perspective of the twentieth century it is easy to see the importance of Tycho's program of systematic observation. It is not so easy to understand why in feudal Denmark he should have been granted the huge sums necessary for equipping and operating his observatories or why he should have been so highly esteemed throughout Europe. Part of the explanation undoubtedly lies in the prevalence of astrology during the sixteenth century. Tycho was a master of the ambiguous "prediction" and had some early luck with royal horoscopes. He and his operation were status symbols as well. His observatory was an architectural marvel, and his instruments were coveted as works of art by more than one

Figure 9.1   Tycho Brahe (1546–1601). Science Museum, London.

foreign prince. Perhaps the enthusiasm for Tycho's projects can be seen as early evidence of a basic drive in post-Renaissance Western civilization, a need to challenge nature, an impatience with mystery. There are parallels between Tycho's efforts and the technology-heavy "big science" of the past twenty years—the giant cyclotrons, linear accelera-

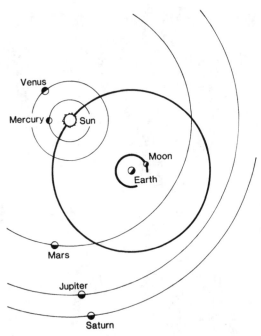

Figure 9.2 The Tychonic system (not to scale). This system is entirely equivalent to the Copernican system, in the sense that the *relative* motions of all the celestial bodies (except the stars) are the same on the two systems.

tors, and radio telescopes funded by a public that understands neither the function of the apparatus nor the goals of the researchers.

In his relations with his tenants, Tycho was autocratic and intolerant, a result perhaps of his upbringing. Born into the Danish royalty, he was kidnapped by a childless uncle who wished an heir to his considerable properties. Tycho was educated at Copenhagen and at German universities. His interests in mathematics and astronomy were at first tolerated and then indulged while he also studied medicine and law. When his uncle died, Tycho, still a student, was free to devote himself fully to astronomy. He studied further in Switzerland and began to commission from the German master craftsmen instruments to take back to Denmark. His reputation was established with the publication in 1573 of *de Nova Stella,* a discussion of the supernova of the previous year. Tycho was able to show unequivocally that the nova did not move among the stars. This was important evidence against the Aristotelian immutability of the heavens.

In Denmark Tycho missed the stimulation he had found in the Eu-

Figure 9.3 One of Tycho's observatories, Stjarneborg (Castle of the Stars), with several kinds of instruments visible. Stjarneborg and Uranienborg, both on the island of Ven, were the most advanced pre-telescopic observatories. From *Tycho Brahe's Atlas Major,* vol. 1, Jan Blaen, 1664. Photo, Science Museum, London.

ropean capitals. In 1576 he was preparing to emigrate to Germany when King Frederick II made him an offer he couldn't refuse. The entire island of Ven, with its lands and forests, was to be at Tycho's disposal for establishing an observatory. During the succeeding years generous grants of other lands and incomes were made. On Ven Tycho built two buildings—the castle, Uranienborg, and the observatory, Stjarneborg (fig. 9.3). Among the instruments he brought there or had built were a great quadrant having a 19-foot radius, the 5-foot diameter "Augsburg sphere," and the famous "mural quadrant (fig. 9.4). At Uranienborg alone there were more than 20 significant astronomical instruments (fig. 9.5). The life at Uranienborg was lusty, but the measurements always came first. Tycho pushed himself as hard as he did his assistants, moving from station to station, checking measurements, sometimes seeming not to need sleep, sometimes napping at one of the cots placed near the larger instruments. The output from Ven was appropriately large:

Figure 9.4 The great mural quadrant at Uranienborg, 1587. The observer is at right
center, while light enters from the slit window at the upper left. Several instruments are
shown in the alcoves in the background. From Atlas Major, vol. 1. British Museum.

almost a thousand stars were precisely catalogued and their positions
inscribed on the great brass sphere; the planets were traced with an
accuracy of 2 to 3 minutes of arc; comets were painstakingly tracked
and eventually found to be beyond the moon; irregularities that Copernicus had labored to explain were shown to be nonexistent. The
stage was being set for the discovery of the true motions of the planets.

Figure 9.5 Tycho's equatorial armillary with a 9½ foot diameter declination circle. From Tycho's *Astronomiae instauratae mechanica*, 1598. Science Museum, London.

In 1596, Tycho's arrogance, his extravagance, and his exploitation of the residents of Ven had alienated the new king, young Christian IV, who might otherwise have been inclined to continue the support his father had begun. Unwilling either to defer to the young king or to adjust his feudal lifestyle, Tycho packed many of his instruments and left Denmark in 1597. He was welcomed to the court of the German Emperor, Rudolph II, near Prague, who had great respect for the astrological skills of the great scientist. Some of Tycho's instruments were still en route when he died in 1601 of complications following an evening of particularly heavy drinking.

Tycho and Kepler

During his brief period at Prague, Tycho added to his staff a highly regarded young German mathematician, Johannes Kepler. Given their personalities, their relationship was not likely to be placid. In *The Watershed,* his biography of Kepler, Arthur Koestler characterizes their meeting:

At last, then, on 4 February, 1600, Tycho de Brahe and Johannes Keplerus, co-founders of a new universe, met face to face, silver nose to scabby cheek. Tycho was fifty-three, Kepler twenty-nine. Tycho was an aristocrat, Kepler a plebian; Tycho a Croesus, Kepler a church-mouse; Tycho a Great Dane, Kepler a mangy mongrel. They were opposites in every respect but one: the irritable, choleric disposition they shared. The result was constant friction, flaring into heated quarrels, followed by half-hearted reconciliations.[3]

Kepler soon established himself as the most skilled of Tycho's computational assistants. As a result he was given the problem of computing the orbit of Mars (around the sun, but in the Tychonic system), which had baffled Longomontanus, Tycho's senior mathematician.[4] Of the six planets, Mars and Mercury have orbits that are farthest from being circular. Because of its nearness to the sun Mercury's positions were less accurately known, and so Mars was the best candidate for discovering the true orbital shape. Indeed, given the level of accuracy of Tycho's data, Mars was the only planet for which epicyclic calculations should have failed. Kepler made little progress during the year he worked under Tycho; for one thing Tycho refused to allow him full access to the collected observations. Kepler was a dedicated Copernican, and Tycho was determined to preserve the Tychonic system. On his deathbed Tycho asked that his system not be abandoned, but Kepler gave no promise.

Kepler was appointed to succeed Tycho as imperial mathematicus to Rudolph II. With characteristic bad luck, Kepler collected practically none of his salary from the Bohemian treasury. There were quarrels with Tycho's heirs over the disposition of the data and the preparation of the full planetary tables based on them. No one but Kepler was qualified to carry out this great labor, and Kepler did not conceal his intention to use the Copernican system. As a result of this conflict and of the vicissitudes of Kepler's middle years, the Rudolphine Tables, as they were named, were not published until 1625. During the more than twenty years that intervened, Kepler published several books based on his calculations from Tycho's numbers.

Kepler

Of all the men who had major roles in the Copernican revolution, Johannes Kepler is surely the most difficult to pigeonhole. Copernicus and Brahe clearly were tied to Renaissance throught patterns. Galileo and Newton—though complex personalities—were modern in their confidence in scientific rationalism. Kepler's preoccupations, on the other hand, were a contradictory mixture of uncompromising mathematical astronomy and the murkiest of analogical mysticism. Koestler's *The Sleepwalkers* derives its title principally from this paradox in Kepler's character. Not dispassionate rational inquiry, but dreams of a geometrical symmetry motivated Kepler, so that his most fundamental discoveries gave him little satisfaction, and his life was full of anticlimaxes and frustrations.

An unusually full description of Kepler's family and his early years survives from an elaborate astrological preparation he made while in his twenties. By his own account, his life was miserable. His family was poor, and he was a sickly child with poor eyesight, who felt himself to be hated by his schoolmates. He was saved from obscurity by the excellent competitive public school system of Württemberg, which assured the education of bright young Protestant men. Neither his poor health nor his misfortunes disappeared in his adult life. His first teaching position was in the Austrian city of Graz, which was predominantly Catholic. At Graz, and indeed throughout his life he was victimized by Counter-Reformation persecution. As a scientist, however, he was treated less severely than the typical Protestant in Catholic principalities. Apart from his work, it seems, he found virtually no satisfaction in his entire life. He made a poor marriage; three of the five children from this marriage did not survive, and his wife Barbara died at 37. He was unable to obtain a position at one of the Protestant universities near his birthplace, and the jobs he did get paid poorly. There were always problems with money. We have remarked that he was unable to collect his wages from Rudolph II; later he expended much time and effort in unsuccessful attempts to secure his second wife's inheritance. In 1620

Figure 9.6 Johannes Kepler (1571–1630). From an engraving by Jakob von Heyden, 1620. Deutches Museum, Munich.

his mother was prosecuted as a witch. It required several months of his time to obtain her acquittal.

Kepler was so little convinced of the importance of his planetary laws that he never summarized them. They are found buried in his *Astronomia Nova* (1609), which grew out of his "war with Mars," and in his *Harmonice Mundi* (1619), which is mostly devoted to an attempt to resurrect the Pythagorean music of the spheres. These books and the several others like them Kepler produced, are the only stream-of-consciousness astronomical treatises ever published. Every step of derivation, every arithmetical mistake, every blind alley, every moment of elation or dejection are described. Although this is of great interest to the historian of science, none of his major works have been translated into English. The story of Kepler's discovery of the laws that bear his name is told elsewhere;[5] we content ourselves with explaining them in modern terms and discussing the context in which they were discovered (fig. 9.7).

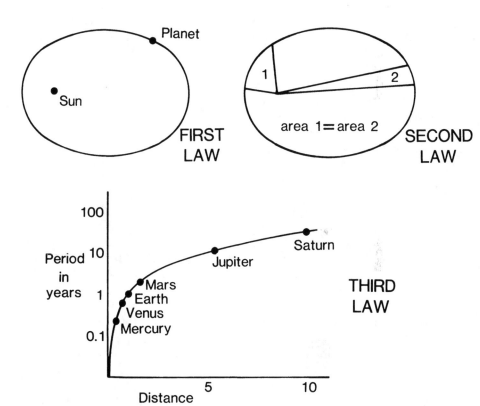

Figure 9.7   Kepler's Laws of Planetary Motion. The First Law is the law of the ellipses. The orbit of a planet is an ellipse, with the sun at one focus. The Second Law is the law of equal areas. The line joining a planet with the Sun sweeps out equal areas in equal times. The Third Law, the harmonic law. The squares of the periods of the planets are proportional to the cubes of their mean distances from the Sun. This can be written $T^2 = kd^3$ or $T = k'd^{3/2}$. In the solar system, with T in years and distance in astronomical units (as in the figure), $T^2 = d^3$.

First Law.   Each planet moves in an orbit that is an ellipse, with the sun at one focus of the ellipse.
Second Law. A line drawn from the sun to a planet sweeps out equal areas in equal times.
Third Law.   The periods of the planets increase uniformly with increasing distance from the sun, the square of the period being proportional to the cube of the planet's mean distance from the sun.

The first law gives the shape of the planetary orbits; the second specifies how the planet moves everywhere on its elliptical path. Any doubt about the essential role of the sun is removed by these laws. When the laws are stated as we just have, it would be hard not to think that in some sense the sun was the cause of the planetary motions. A satisfactory explanation of the causal connection between the sun and the planets was to come from Newton, but Kepler did have some ideas about solar forces. Impressed by William Gilbert's investigations into magnetism, Kepler thought that the sun was like a large, rotating magnet that drew the planets around like so many floating iron needles. This would qualitatively account for the second and third laws, since the impelling force would be weaker for greater separations. It is significant that in these speculations Kepler was concerned with dynamics, which for two thousand years had not been an element of astronomy.

The first two of Kepler's laws came out of his four-year struggle with the orbit of Mars. Kepler was not content to predict the angular positions of that planet. He also required that the distances from Mars to the sun be correctly given; these distances were obtained by triangulation, not by observing the variable brightness of Mars. The problem was simplified early in the battle when Kepler verified that the plane of the orbit of Mars is at a constant angle with respect to the ecliptic and passes through the sun. The latitudes of the planet were then trivially derivable from the longitudes, and Kepler could concentrate on the longitudes and the distances. He found an excellent fit to the longitudes with a simple eccentric-plus-equant construction. But the distances were not explained by this construction. The error was small; the best fit to the distances forced a maximum error of only 8 minutes (one-eighth degree) in the angular positions. It was at this point that Kepler abandoned circular orbits. It is interesting that Kepler was already influenced by his belief that the sun should be at a physically meaningful place,[6] whereas the most significant point of the circular orbit was the equant, a point in empty space. He first tried adding an epicycle and finally abandoned compounded circular orbits entirely. After attempting various oval shapes, including an egg shape, he finally discovered the ellipse and the first law. He had already in the course of the calculation discovered the second law, which correlates the distance from the sun with the planet's velocity.[7] Because there was no mathematics to deal efficiently with the very small quantities involved in the calculations, Kepler was required to work by trial and error, repeating elaborate calculations dozens of times. His work was made even more tedious by arithmetic errors. He was entirely scrupulous about the accuracy of his calculations, however, several times rejecting an apparently satisfactory solution when an error was discovered.

Kepler's third law ($T \propto d^{3/2}$) is a quantification of a regularity that had been assumed for a very long time, namely, that the more distant planets move more slowly. Copernicus had in fact referred to the qualitative relationship—that Saturn moves most slowly, Jupiter next slowly, etc.—as the "first law" of planetary motion. The third law was Kepler's favorite, because it dealt not with a single planet but with the entire "world." Since his youth Kepler had been seeking a geometrical confirmation of a unifying harmony among the planets. While still in his twenties he had discovered, or thought he had, such a unity revealed in the relative distances of the planets. There are six planets, so there are five spaces between the planets. Kepler recalled that there are just five regular polyhedrons, and he proposed a scheme of alternating the polyhedra with inscribed spheres (fig. 9.8), choosing the order of the regular solids so as to obtain best agreement with the observed distances. For the sequence octahedron, icosahedron, dodecahedron, tetrahedron, cube (reading from the inside out), the agreement was fair. Kepler never really gave up on this idea, that the locations of the planets have some fundamental significance. He tried a number of other schemes, but he never succeeded in obtaining good agreement.

Perhaps the most fantastic of Kepler's attempts to find unifying numerical relations is found in the *Harmonice Mundi,* where he literally harmonizes the universe. The Pythagoreans had, before the fourth century B.C. proposed that each planet produces a musical tone as it circles in the heavens. Kepler observed that if this were true, the sounds must not be constant in pitch, since the planets vary in speed as they move (in ellipses, as he already knew) about the sun.

Kepler, then, is an enigmatic figure, the least well understood of the major founders of modern science. When we try to make a modern figure of him, his mysticism, his search for celestial harmony, and his vision of the sun as the abode of God rudely remind us how medieval in spirit he sometimes was. But if we try to dismiss Kepler as mired in astrology, sun worship and hermeticism, we need only to recall his

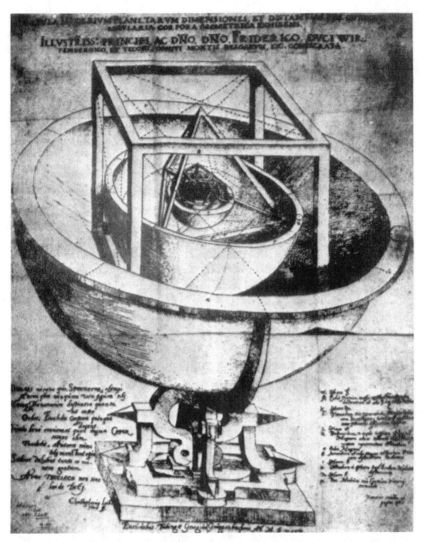

Figure 9.8 The nested regular solids. The cube is inscribed within the sphere of Saturn's orbit; the sphere of Jupiter's orbit is inscribed in the cube. A tetrahedron is inscribed within Jupiter's sphere; Mars' sphere is inscribed in the tetrahedron, and so on. From Kepler's *Mysterium Cosmographicum*. Reprinted with permission of Pergamon Press, Ltd. and Owen Gingerich.

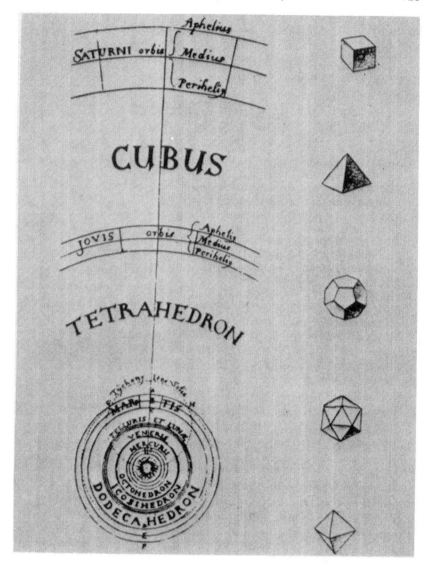

contributions to planetary dynamics and optics, his founding, with Galileo, of the new astronomy. Much work needs to be done in filling out the figure of Johannes Kepler, to make him believable to this century. Above all, we desperately need good translations into English of his major works.

CHAPTER 10

Galileo Galilei

And new Philosophy calls all in doubt,
The Element of fire is quite put out;
The Sun is lost, and the earth, and no man's wit
Can well direct him where to looke for it.
And freely men confesse that this world's spent,
When in the Planets, and the Firmament
They seeke so many new; then see that this
Is crumbled out againe to his Atomies.
'Tis all in peeces, all cohaerence gone;
All just supply, and all Relation:
Prince, Subject, Father, Sonne, are things forgot,
For every man alone thinkes he hath got
To be a Phoenix, and that then can bee
None of that kinde, of which he is, but hae.

                              An Anatomie of the World,
                              —John Donne (1611)

In 1564, the year of the death of Michelangelo and the birth of William Shakespeare, Galileo Galilei (fig. 10.1) was born. It can be said about this great intellect and extraordinarily rich and human personality that in his lifetime the transition from medieval to modern science occurred. Through his discoveries the distinction between heaven and earth was largely wiped out and Aristotelianism was dealt a death blow. Although he did not introduce experimentation into physics, he was the first to show its force and to effectively weld the superficially opposing principles of experimentation and mathematical reasoning into a unified scientific methodology. By isolating the essential elements of a real experiment and through a process of abstraction deducing the outcome of the idealized unperformable experiment, he showed the power of what we now call the "thought experiment." His attempts to express observed regularities in terms of a mathematical abstraction set mathe-

Figure 10.1 Galileo Galilei (1564–1642), by Ottavio Leoni, 1624. Courtesy of Stillman Drake.

matical physics on its way. A foe of naïve generalization from experiment, he showed the crucial role of carefully planned experimentation based on prior theory, as opposed to simple observation, and of the predictive value of mathematical theory. By the end of his life he was a thoroughgoing mathematical physicist, though conscious of the importance of experiment in verification and falsification. He spent as much of his long life in advocacy, in persuasion, as in experimentation and theorizing. He accepted the burden of defending the right of science to claim authority where questions of scientific knowledge were concerned, a conflict in which he spent more than ten years of his life. Galileo was vital, argumentative, neither martyr nor coward. He was shrewd and practical but a typical Renaissance humanist, a lover of Italian poetry and music. Of all the founders of modern science, Galileo comes down to us as the most sympathetic, the most believable.

Galileo was born in Pisa, where he received his early education from Jesuit scholars. Though his father urged upon him a career in medicine, in 1583 he discovered mathematics and the works of Archimedes and soon abandoned his earlier studies. His first academic position was professor of mathematics at the University of Pisa, from 1589 to 1592. It was there that he is supposed to have carried out his famous demonstration that the descent of a falling body is independent of its weight. Whether the story is true or not is not very important;[1] Galileo clearly knew what the result of such an experiment would be and at the very least discussed it as a thought experiment. In 1592 Galileo took a position at the University of Padua, which he held for eighteen years. This period, perhaps the happiest of his life, was extraordinarily fruitful; the Venetian republic provided an intellectual climate and a freedom that played a crucial role in his development. It was here that he performed his early studies of motion.

In 1609, having heard from Holland of an invention that brought distant objects nearer, he constructed his first telescope and soon had perfected a model that magnified 30 times (fig. 10.2). By 1610 he had made a number of remarkable astronomical discoveries, some having the most far-reaching philosophical consequences. That year Galileo published his *Starry Messenger* (*Siderius Nuncius*), in which his earliest telescopic discoveries were described. He discovered the mountainous nature of the moon's surface, making it clear that it was not entirely unlike the earth, and observed the four major satellites of Jupiter for the first time. Both of these observations raised serious problems for the Aristotelians. His discovery that the stars still appeared as points even when magnified thirty times destroyed one of Tycho's arguments against the Copernican system. Long a Copernican (having supported the Copernican theory in a 1597 letter to Kepler upon receipt of a copy of *Mysterium Cosmographicum*), Galileo declared his unqualified support of the heliocentric system in his "Letters on Sunspots" of 1613.[2]

In the meantime, Galileo had taken advantage of the acceptance of the *Starry Messenger* to gain a position in Florence in 1610. This move was eventually to have fateful consequences because of the weakness of

Figure 10.2 Two of Galileo's telescopes on a mounting which includes a broken objective lens. Museo di Storia della Scienza, Florence.

his patron, Cosimo dé Medici, in the face of demands from Rome. By 1614 Galileo's support of the theory of the moving earth had gained some attention, and when in 1616 Copernicus' *De Revolutionibus* was banned, Galileo was warned by Jesuit Cardinal Bellarmine to observe the general injunction against adherence to the Copernican doctrine that had been issued in that year. Rumors that he had been forced to abjure induced Galileo to secure from Bellarmine a statement to the contrary. This was to become for Galileo an important defense against the Inquisition. In this episode of 1616, Galileo had been pressed mainly by influential Dominican prelates but was held in high esteem by many of the Jesuits, including Bellarmine and Cardinal Maffeo Barberini, eventually Pope Urban VIII. His relationship with the Jesuits was to change drastically during the next sixteen years, in part because of the controversy that arose with Jesuit Father Scheiner over priority in the matter of the discovery of sunspots and for a similar controversy over comets with Father Grassi.

The latter dispute prompted Galileo to write *The Assayer* (*Il Saggiatore*) his "scientific manifesto." In 1624 Galileo appealed in Rome to the newly installed Pope Urban VIII, to be allowed to publish his ideas on the systems of the world. Permission was granted, with the restriction that the rival theories be described impartially. The work, *Dialogue Concerning the Two Chief World Systems,* was published in 1632. Thus began the inexorable process that was to bring Galileo to his knees before the officers of the Inquisition. The work was shortly withdrawn and banned and Galileo ordered to trial in Rome. Nonetheless, the *Dialogue* was published in Holland in 1635 by Elsevier, and knowledge of it spread widely across Europe. It is clearly a Copernican treatise, containing detailed arguments in favor of the heliocentric system.

Following his trial and abjuration, which we explore later, Galileo soon set to work again, gradually recovering from discouragement over his disgrace and imprisonment by the Church. By 1634 he was completing a new work, whose origins went back more than 20 years, to be published in 1638 as the *Two New Sciences*. This work, his scientific masterpiece, is devoid of discussion of world systems, as it had to be under the circumstances. In 1638 Galileo became totally blind, and the remainder of his life was spent with students, including Vincenzio Viviani and Evangelista Torricelli,[3] and his son Vincenzio, and in a wide-ranging scientific correspondence. He died at Arcetri near Florence on January 9, 1642, in the year of the birth of Isaac Newton, who would take Galileo's mechanics and build upon it the immense edifice of classical physics. He is buried in the Church of Santa Croce in Rome, close to the tomb of Michelangelo.

Galileo and the Church

The confrontation between Galileo and the church is one of the most extraordinary moments in the history of science. A single bold spirit sought to overthrow the established scientific dogma of the church and to free science from the shackles of religious doctrine. It is a tragedy not only for the protagonist, who was forced ultimately to deny the work of a lifetime, but also for all mankind, and particularly for those who could delude themselves with the thought than an idea could be killed by crushing the spirit of one man. The tragedy, then, ends as it must, in triumph. For no power, secular or sacred, could stop the advance of the scientific spirit of which Galileo is the embodiment.

Galileo was a loyal, if not particularly devoted, Catholic. His goal was not confrontation with the church but rather its enlightenment. He attacked the Aristotelian basis of Catholic theology, not from an interest in church doctrine per se (he deferred to the judgment of the church in purely religious matters), but rather because of the restraining effect of Aristotelian metaphysics on studies of motion and of the motion of the earth in particular. The difficulties that inevitably arose from this program led to Galileo's persistent and ultimately tragic efforts to establish the preeminence of science in elucidating the natural world.

Galileo's astronomical discoveries and his teachings on the Copernican theory aroused the suspicion of conservative scholars almost from the very beginning. The reasons were several but centered upon Galileo's criticism of Aristotle and on the supposed conflict between the heliocentric theory and the scriptures. Yet many church scholars, particularly Jesuit theologians,[4] were immensely interested in Galileo's discoveries, and in 1611 Galileo journeyed to Rome to explain these discoveries to the church hierarchy. His reception was favorable, and the Jesuits for the most part accepted his observations, even though they disputed his interpretation of them. Far more open to discoveries in the sciences at the time than, for example, the Dominicans, the Jesuits nonetheless were the traditional guardians of dogma and orthodoxy and were quick to sense the difficulties that might arise from the novel observations of Galileo. As early as 1611 Cardinal Bellarmine was troubled by the attention given to the telescope and was closely following Galileo's activities. Although the Jesuits were not of one mind on the matter of Copernicus up to 1616, after that date they generally retreated to the safety of the Tychonic system. A quotation from a 1615 letter of Bellarmine to Foscarini, who had tried to reconcile the Copernican system with scripture, is indicative of the climate of the period:

I say that it appears that your Reverence and Signor Galileo did prudently to content yourselves with speaking hypothetically and not positively, as I have always believed Copernicus did. For to say that assuming the earth moves and the sun stands still saves all the appearances better than eccentrics and epicycles is to speak well. This has no danger in it, and it suffices for mathematicians.

But to wish to affirm that the sun is really fixed in the center of the heavens and merely turns upon itself without traveling from east to west, and that the earth is situated in the third sphere and revolves\very swiftly around the sun, is a very dangerous thing, not only because it irritates all the theologians and scholastic philosophers, but also because it injures our holy faith and makes the sacred Scripture false. . . .

I say that if there were a true demonstration that the sun was in the center of the universe and the earth in third sphere, and that the sun did not go around the earth but the earth went around the sun, then it would be necessary to use careful consideration in explaining the Scriptures that seemed contrary, and we should rather have to say that we do not understand them than say that something is false which has been proven.[5]

Attacks against Galileo began as early as 1612. In late 1615 he returned to Rome, in an attempt to explain the new theory and to clear his name of any suspicion of heresy. In fact Galileo was there at considerable personal risk and had many enemies, and the Pope, Paul V, was inclined to invoke the machinery of the Inquisition. The Florentine ambassador, Guicciordini, wrote:

I hear that Galileo is coming. . . . When I first came here [in 1611] he was here, and spent some days in this house. His views, and something else too, did not please the advisors and cardinals of the Holy Office. Among others, Bellarmine told me that . . . if he was here too long, nothing less could be done than to arrive at some judgment concerning his affairs. . . . I do not know if he has changed his views or his temper, but I know very well that some Dominicans who play a great part in the Holy Office, as well as others, are ill-disposed toward him, and this is no place to come to argue about the moon, nor, in this age, to support or import any new doctrines.[6]

Shortly after Galileo's arrival in Rome, the Holy Office took up consideration of the propositions (1) That the sun is the center of the universe, and consequently is not moved by any local motion; and (2) That the earth is not the center of the universe nor is it motionless, but moves as a whole, and also with the diurnal motion. On February 24, 1616, the first proposition was declared "formally heretical" and the second "erroneous in faith." The works of Copernicus were banned, as were all other books teaching the same doctrine. As for Galileo, Bellarmine was instructed by the Pope to admonish him to abandon the censured views, and on February 26, with the Commissary General of the Holy Office evidently present, Bellarmine so warned Galileo. The question of whether the examination went further, to an order not to hold, teach, or defend the Copernican doctrine "in any way," is still unresolved and plays a central role in the trial of 1633. In any event, Galileo obtained from Bellarmine the following statement:

We, Robert Cardinal Bellarmine, having heard that Signor Galileo Galilei is calumniated or imputed to have abjured in our hand, and even of having been given salutary penance for this; and inquiries having been made as to the truth, we say that the said Signor Galilei has not abjured any opinion or doctrine of his in our hand nor in that of anyone else at Rome, much less anywhere else, to our knowledge; nor has he received penance of any sort; but he has only been told the decision made by His Holiness and published by the Holy Congregation of the Index, in which it is declared that the doctrine attributed to Copernicus, that the earth moves round the sun and that the sun is fixed in the center of the universe without moving from west to east, is contrary to the Holy Scriptures, and therefore cannot be defended or held.[7]

After the election of Cardinal Barberini to the papal chair in 1623 as Urban VIII, Galileo decided to again test the mood of the church on the matter of the Copernican theory. In that year he published *The Assayer* and dedicated it to Urban, who was said to be much pleased by the dedication. Then in April 1624, Galileo went to Rome, where he was warmly received by the Pope. Although his pleas to have the 1616 decree revoked were met with evasion, Galileo received the impression that debate over the theory would not be opposed.

Thus in 1624 Galileo began his great *Dialogue on the Two Chief World Systems,* originally intended to be named *Dialogue on the Ebb and Flow of the Tides* but retitled at the demand of the censors because of Galileo's well-known insistence that his theory of the tides provided a conclusive proof of the heliocentric theory. The *Dialogue* was finished in 1630 and, after much delay caused by pressure from Galileo's enemies, was reluctantly given the imprimatur of the church and published in Florence in February 1632. Almost immediately the book was condemned, and by October the order to stop its sale and to retrieve all copies was issued. Urban was enraged, having been convinced that Galileo had not only advocated the Copernican theory against the Ptolemaic but had also deceived him by not informing him of the injunction alleged to have been delivered to him by the Commissary General in 1616. Moreover, the Pope may have been persuaded that the slightly dim-witted peripatetic of *Dialogue,* Simplicio, was modeled after the Pope himself.

External circumstances no doubt influenced Urban's reaction to the publication of the *Dialogue.* If his position in the Catholic world had been more secure, he might have felt less strongly the need to react unequivocally against what he saw as an attack on his authority and that of the church. The Pope was under pressure from both Spain and France, each of whom wanted him to intervene in their political struggle, and in Rome faced accusations of nepotism. Thus his own troubles demanded a vigorous and decisive response to Galileo's threat to orthodoxy. It is also likely that the opposition of the Jesuits to Galileo, particularly Father Scheiner, who is thought by some to have been instrumental in having Galileo condemned, was important, as well as tensions between the Dominicans and the Jesuits. In any case, Galileo was directed to appear before the Commissary General of the Inquisition, and after exhausting all appeals he presented himself in Rome in February 1633.

Galileo approached his trial with trepidation but confident that he could convince his accusers of the validity of his arguments. He had devoted much of the previous 22 years at Florence to advocating the Copernican theory and the role of science in providing absolute knowledge of the natural world and in establishing the primacy of reason over religion in matters of science. In fact he was given no chance to defend his ideas; rather, he found himself accused of violating the 1616 decree and of failing to inform the censor of the prohibition supposedly delivered to him by the Inquisition at that time. He was confronted with the charge of teaching that the Copernican theory is consistent with scripture, a position that had been declared heretical and one that he had been warned, allegedly, not to hold. The actual trial began on April 13, finally concluding in June with Galileo's abjuration. The charges against Galileo were based on an unsigned minute of the 1616 interview, wherein Galileo was, it was claimed, ordered not to hold, defend, or teach in any manner, the Copernican doctrine. Most scholars have doubts about the validity of the document or of the interview itself,[8] especially in view of the certificate provided by Bellarmine to Galileo.[9] Nonetheless, it provided the basis for the case against Galileo.

Faced with the hopelessness of his cause, Galileo acknowledged having unintentionally favored the Copernican theory in the *Dialogue* out of vanity and the cleverness of his arguments. There is good reason to believe that an understanding had been reached that this "confession" was to be exchanged for a light sentence.[10] Apparently, the agreement was repudiated by the Pope. So on May 10, Galileo was told to write out his defense, and on June 21 he appeared for "rigorous examination," which included the possibility of torture. Having lost all heart for further struggle, 69 years old and in poor health, facing excommunication, perhaps even torture and death, Galileo renounced his life's work.

I do not hold nor have I held to this opinion of Copernicus since the precept was given to me that I must abandon it; for the rest I am in your hands, and you may do as you please.[11]

The sentence of imprisonment "at the pleasure of the Holy Office" was delivered on June 22:

We say, pronounce, sentence, declare that you, the said Galileo by reason of the matter adduced in trial, and by you confessed as above, have rendered yourself in the judgment of this Holy Office vehemently suspected of heresy, namely of having believed and held the doctrine—which is false and contrary to the sacred and divine Scriptures—that the Sun is the center of the world and does not move from east to west, and that the Earth moves and is not the center of the world; and that an opinion may be held and defended as probable after it has been declared and defined to be contrary to Holy Scripture; and that consequently you have incurred all the censures and penalties imposed and promulgated in the sacred canons and other constitutions, general and particular, against such delinquents. From which we are content that you be absolved, provided that first, with a sincere heart, and unfeigned faith, you abjure, curse, and detest the aforesaid errors and heresies, and every other error and heresy contrary to the Catholic and Apostolic Roman Church in the form to be prescribed by us.[12]

Following the reading of the sentence, Galileo was forced to publicly abjure:

because, after I had received a precept which was lawfully given to me that I must wholly forsake the false opinion that the sun is the center of the world and moves not, and that the earth is not the center of the world and moves, and that I might not hold, defend, or teach the said false doctrine in any manner, either orally or in writing, and after I had been notified that the said teaching is contrary to the Holy Scripture, I wrote and published a book in which the said condemned doctrine was treated, and gave very effective reasons in favor of it without suggesting any solution, I am by this Holy Office judged vehemently suspect of heresy, that is, of having held and believed that the sun is the center of the world and immovable, and that the earth is not its center and moved.

. . . wishing to remove from the minds of your Eminences and of every true Christian this vehement suspicion justly cast upon me, with sincere heart and unfeigned faith I do abjure, damn, and detest the said errors and heresies, and generally each and every other error, heresy, and sect contrary to the Holy Church; and I do swear for the future that I shall never again speak or assert, orally or in writing, such things as might bring me under similar suspicion. . . .[13]

Three cardinals of the ten who tried Galileo, including Francesco Barberini, nephew of the Pope, refused to sign the sentence. The life sentence was swiftly reduced by Cardinal Barberini to confinement at the

Palace of the Florentine ambassador, and then at Siena. By the end of 1633 he was allowed to return to his villa at Arcetri.

The abjuration, the "depth of abasement" to some, was inescapable from the very start. When faced with separation from his church and his God, through excommunication, Galileo, a good Catholic, could do no other. That he was confronted with possible torture or death[14] should not be forgotten, but it seems most reasonable to attribute Galileo's final renunciation of his deepest beliefs to his inability to continue to oppose the church when he came to understand that he could no longer hope to prevail against it. In 1633 he wrote:

I do not hope for any relief, and that is because I have committed no crime. I might hope for and obtain pardon, if I had erred; for it is to faults that the prince can bring indulgence, whereas against one wrongfully sentenced while he was innocent, it is expedient, in order to put up a show of strict lawfulness, to uphold rigor.[15]

Did Galileo say, following his trial and abjuration, *Eppur si muove,* "still it moves" as legend has for so long held? Certainly not at the trial, for that would have been fatal. But later, quite possibly so. In any case, we know that he continued to believe in the Copernican system, even though he had to remain silent.

Galileo as Polemicist

Galileo, as we have already seen, spent much of his life in advocacy and polemic. Not only did he undertake to defend and explain the Copernican system with vigor and clarity, but also, beyond that, he forcefully argued the role of science in discovering truth in the natural world. From 1604 until the decree of 1616 he taught and wrote on the Copernican theory, using some of his astronomical discoveries, including that of sunspots,[16] to support his arguments. The opposition he encountered, principally from those who worried about the impact on the Aristotelian roots of Catholic theology and those who thought the Copernican system in conflict with scripture, led to a widening of the scope of his arguments, to the advocacy of the primacy of human reason in matters of science. He acknowledged the authority of the church in questions of ethics and religion but denied it any right to pass on scientific questions. To Galileo there was a single truth, and when scientific demonstration and religious dogma collided, the latter had to yield.

The clearest expression of these ideas is to be found in his remarkable *Letter to the Grand Duchess Cristina (1615):*

I think that in discussions of physical problems we ought to begin not from the authority of scriptural passages, but from sense-experiences and necessary demonstrations; for the Holy Bible and the phenomena of nature proceed alike from the divine Word, the former as the dictate of the Holy Ghost and the latter as the observant executrix of God's commands. It is necessary for the Bible, in order to be accommodated to the understanding of every man, to speak many things which appear to differ from the absolute truth so far as the bare meaning of the words is concerned. But Nature, on the other hand, is inexorable and immutable; she never transgresses the laws imposed upon her, or cares a whit whether her abstruse reasons and methods of operation are understandable to men. For that reason it appears that nothing physical which sense-experience sets before our eyes, or which necessary demonstrations prove to us, ought to be called in question (much less condemned) upon the testimony of biblical passages which may have some different meaning beneath their words.[17]

Between 1616 and 1623 Galileo was largely silent, in deference to the decree of the Holy Office, but in 1624, following Barberini's ascent to the papacy, which raised the hopes of progressive Catholics that a period of tolerance in science and the arts was to begin, Galileo published *The Assayer,* a masterpiece of polemic literature, and only incidentally concerned with the problem of comets. A treasure of scientific methodology, philosophy, and anti-Aristotelian polemic, *The Assayer* better than any other work expressed Galileo's understanding of the nature of the search for scientific truth.

After this, Galileo felt encouraged to again attempt to raise with church scholars and officials the matter of the Copernican theory. This time he undertook the more limited goal of dissuading the church from continued repression of the theory. Following the visit to Rome in 1624, he began work on the *Dialogue,* his "Copernican manifesto," one of the monumental works of science, finally completed in early 1630 (fig. 10.3). Superficially the *Dialogue* strikes a balance between the Ptolemaic and Copernican systems and thus gives lip service to the dictates of the Pope. And yet the work is thoroughly Copernican, with the Aristotelian arguments against the earth's motion clearly disposed of. It concludes with Galileo's tidal theory, which the great scientist believed provided a conclusive argument in favor of the heliocentric theory.

The *Dialogue* is divided into four days, the first of which is devoted to a critique of Aristotelian physics. The second and third days deal with Aristotle's arguments against the two motions of the earth, and the fourth deals with the tides. This most famous of Galileo's works is,

Figure 10.3 Frontispiece to Galileo's *Dialogue on the Two Great World Systems.* The figures are Aristotle, Ptolemy, and Copernicus. British Museum.

to paraphrase Alexandre Koyre,[18] polemical, pedagogical, and philosophical. Written in Italian, it is addressed to the cultured layman, to which fact we owe its light conversational tone, its digressions, and repetitions. It is full of eloquent discourse and biting irony and satire; it attacks prejudice and authority and demands carefully reasoned argument. It assails not only traditional physics and cosmology but also the entire philosophical *Weltanschauung* of its adversaries. Among the most important and most persuasive parts of the *Dialogue* are the refutations of arguments against the moving earth. One of the foundations of these arguments is the "principle of Galilean relativity."

Shut yourself and a friend below deck in the largest room of a great ship, and have there some flies, butterflies, and similar small flying animals; take along also a large vessel of water with little fish inside it; fit up also a tall vase that shall drip water into another narrow-necked receptacle below. Now, with the ship at rest, observe diligently how those little flying animals go in all directions; you will see the fish wandering indifferently to every part of the vessel, and the falling drops will enter into the receptacle placed below. . . . When you have observed these things, set the ship moving with any speed you like (so long as the motion is uniform and not variable); you will perceive not the slightest change in any of the things named, nor will you be able to determine whether the ship moves or stands still by events pertaining to your person. . . . And if you should ask me the reason for all these effects, I shall tell you now: "Because the general motion of the ship is communicated to the air and everything else contained in it, and is not contrary to their natural tendencies, but is indelibly conserved in them." [19]

In a devastating series of arguments in favor of the revolution and rotation of the earth, Galileo (in the person of Sagredo) concludes:

SAGR.: If in the totality of effects which may in Nature depend upon such like motions, there should follow in one hypothesis exactly all the same consequences as in the other. I would esteem at first inspection, that he who should hold it more rational to make the whole Universe move, in order to keep the Earth from moving, is less reasonable than he who being at the top of the dome of your Cathedral in Florence, in order to behold the city and the fields about it, should desire that the whole country might turn round, so that he might not be put to the trouble to turn his head. And surely the conveniences that could be drawn from this position would have to be many and great in order to equate in my mind, and to overcome, this absurdity in such manner as to make it more credible than the fomer. But perhaps Aristotle, Ptolemy, and Simplicius must find certain advantages therein which they would do well to communicate to us also, if any such there be; or else they had better declare that there neither is nor can be any. [20]

Galileo concluded the *Dialogue* with what is known as "the argument of Urban VIII":

I know that if asked whether God in his infinite power and wisdom could have conferred upon the watery element its observed reciprocating motion using some other means . . . both of you would reply that He could have, and that He

would have known how to do this in many ways which are unthinkable to our
minds. From this I forthwith conclude that, this being so, it would be excessive
boldness for anyone to limit and restrict the Divine power and wisdom to some
particular fancy of his own.[21]

By this argument, unconvincing and incongruous in view of the tenor
of what preceded it, Galileo could argue that he had obeyed the injunc-
tions of the Pope. Yet he fooled no one. The response to the publication of the *Dialogue* was immediate and decisive. The work was banned,
and in just over a year, Galileo was before the Inquisition in Rome.
This meant the collapse of his role as advocate and propagandist. Although he corresponded widely, and was visited by Hobbes and Mil-
ton, he was allowed few visitors and was prohibited from teaching.
Nothing was to be published. For this reason his great scientific document, the *Two New Sciences,* is devoted entirely to statics and dynamics
and was published in Holland. In 1744 the *Dialogue* was allowed to be
republished by the church but with "corrections"; it was not until 1822
that the ban on his works was fully lifted.

Astronomy

Galileo's contributions as an astronomer, while clearly a product of
the same scientific imagination that founded modern mechanics, are
nonetheless of a sufficiently different character from that work as to
justify their analysis apart from their value as observations. In the years
after 1609, Galileo made a series of observations with the telescope that
not only brought to light hitherto unsuspected phenomena but also,
more importantly, greatly weakened the credibility of the Aristotelian
two-sphere universe and provided, moreover, direct support for the
Copernican theory as against the Ptolemaic. Furthermore, Galileo had
the vision to see the telescope as an important tool of science when
others thought it a toy or even a distorter of reality. This was in itself
a contribution of the first order.

By January 1610 Galileo was able to assert that the moon was much
like the earth and that it possessed mountains even higher than those
known on the earth (fig. 10.4). This discovery, when fully appreciated,
was an important step in "making the earth a planet," in that it demonstrated that one celestial body, at least, was very much like the earth
and, moreover, characterized by all sorts of irregularities, mountains,
craters, and the like. Shortly after his observations of the moon, Galileo
discovered the four largest satellites of Jupiter (fig. 10.5), which he

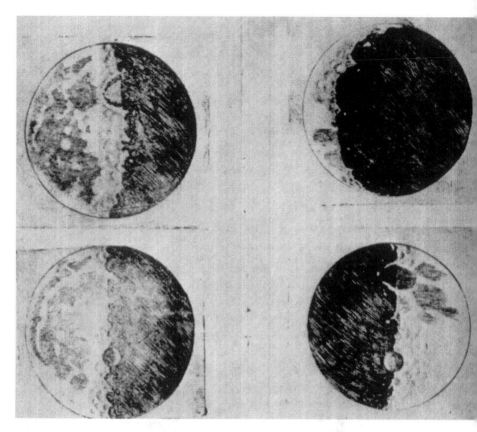

Figure 10.4 Galileo's drawings of the moon. From the *Starry Messenger,* 1610. Biblioteca Nazionale Centrale, Florence.

named "The Medician Planets," but which we now call the "Galilean Satellites." This discovery demolished the a priori arguments about the number of "planets," and Galileo's demonstration that the satellites must revolve about Jupiter assaulted the belief that the planets were carried by impenetrable crystalline spheres. These observations were published in the *Starry Messenger.*

Before leaving for Florence, Galileo discovered sunspots. (These phenomena, however, had been observed in earlier times, perhaps by the earliest astronomers—without any understanding of their character.) [22] He also described the sun's rotation, in his "Letters on Sunspots" in 1613. As we have seen, this embroiled him in a bitter controversy with Scheiner, who may have independently discovered the sunspots. Father

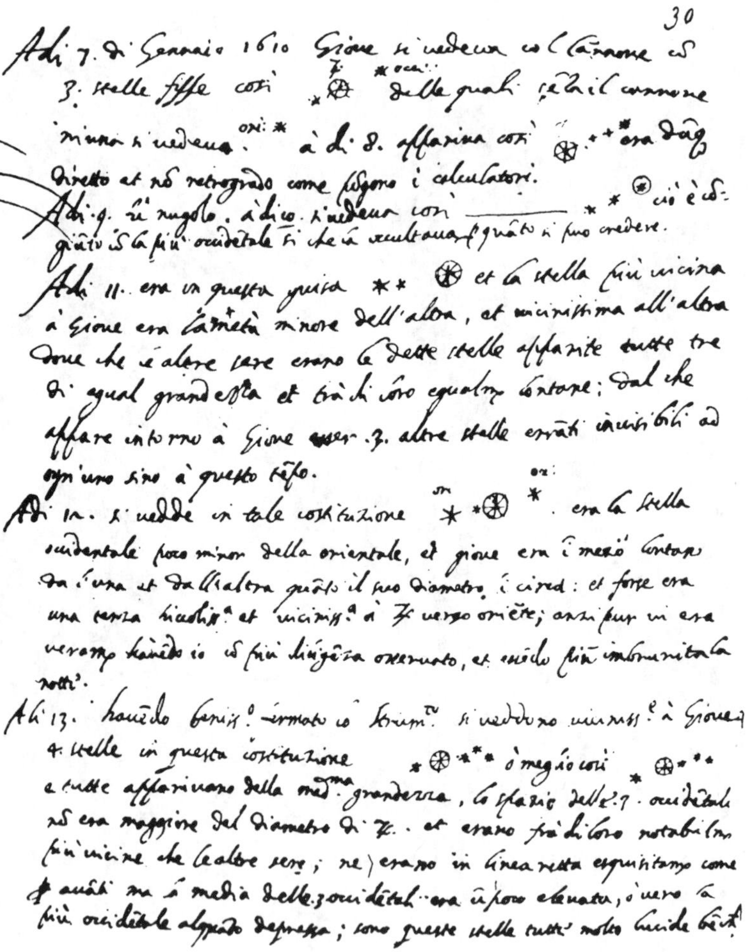

Figure 10.5 A page from Galileo's observational notes on the satellites of Jupiter. Biblioteca Nazionale Centrale, Florence.

Scheiner first believed them to be objects revolving about the sun, rather than surface markings, but eventually he went on to make important and systematic observations of the sun. At about the same time Galileo observed the rings of Saturn, which he saw as two fixed companions of the planet, and the phases of Venus. Both observations were communicated to Kepler in cryptic form. Galileo's observations of Venus were of crucial importance, since it goes through a full range of phases. In the Ptolemaic system this would have been impossible.[23] His obser-

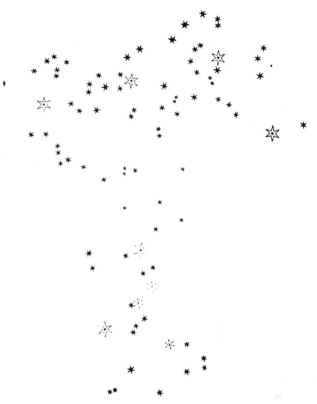

Figure 10.6 Galileo's representation of the stars in the vicinity of the belt and sword of Orion. From *The Sidereal Messenger,* M. T. Cardini, trans. (Florence: Sansoni, 1947).

vation that stellar images show no perceptible disc implied that they were very far distant, as required by the heliocentric theory. Thus Galileo's observations not only opened the era of telescopic astronomy, making it possible to see celestial objects invisible to the unaided eye (fig. 10.6), but also had a profound impact on man's understanding of the universe and on the philosophical background into which the observations had to be absorbed—this in spite of opposition, which dismissed the views exhibited by the telescope as distortions or declared that in any case reality was not to be discovered by looking into such a device, whatever it might show. These arguments, based on a Platonic epistomology or on a religious conservatism that denied any evidence that might undermine the Aristotelian bases of Catholic theology, entirely rejected the telescope as an instrument of science.

It has often been regretted that the two founders of modern astronomy, Kepler and Galileo, failed to achieve any real communication, in
spite of their familiarity with each other's work. But, as Herbert Dingle
has said:

in truth it could scarcely have been otherwise. When one has for the first and
last time discovered the eternal laws laid down by God and the Creation, and
has for the first time and forever opened the ears of the human mind to the
divine harmonies, is he likely to be impressed by a new toy or by the childish
timing of balls running down grooves? And when another has found the keys
of new and unimagined realms of discovery, and sees day by day unfolding
before him fresh wonders and ever-increasing possibilities of knowledge hitherto undreamed of, is he likely to be impressed by the final rounding off of a
played-out theme? That Kepler and Galileo were contemporaries was the mer-
est accident. In temper, outlook and achievements they were centuries apart.[24]

Moreover, as a Protestant, Kepler could be of no help to Galileo in his
difficulties with the church. We may, however, be forgiven the urge to
speculate on how different the course of seventeenth-century astronomy and physics might have been if those two geniuses had understood
each other. That Galileo perceived them as having a common enemy is
indicated by this passage from a letter: "My dear Kepler, what would
you say of the learned here, who replete with the pertinacity of the asp,
have steadfastly refused to cast a glance through the telescope? What
shall we make of all this? Shall we laugh, or shall we cry?"[25]

Galileo and Modern Dynamics

Galileo's early work on motion (*De motu,* written before 1600) is obviously derivative and firmly grounded in medieval impetus mechanics,
with its Aristotelian basis. Between 1602 and 1609 Galileo concerned
himself fruitfully with research into mechanics and by 1604 had formulated the law of free fall, $S \propto t^2$, that the distance traversed is proportional to the square of the time elapsed:

Reconsidering the phenomena of motion, in which I totally lacked any indubitable principle that could be put down as an axiom for the demonstration of
the events I have observed, I hit upon a proposition which contains much of
the natural and evident. This being assumed, I can then demonstrate the rest;
that is, that the spaces passed over in natural motion are in proportion to the
squares of the times, and consequently the spaces passed over in equal times are
as the odd numbers beginning from one; and other things. Now the principle
is this: that the body in natural motion increases its speed in the same proportion as its departure from the origin of its motion.[26]

Galileo was "wrong" about this "indubitable principle," for the square of the speed is proportional to distance traveled,[27] but his relation between distance and time was correct.

Galileo's interest in mechanics never waned, but during the next 25 years, he devoted much of his time to the Copernican theory and his program of scientific advocacy. When he began to write the *Two New Sciences* in 1633–34 he was reworking and revising work begun during the Paduan period.

The *Two New Sciences* is divided into two parts comprising four "days." The first two deal with statics, particularly the strength of materials, while the final two develop the science of dynamics. Fragments of proposed continuations are also known. The laws of uniform and uniformly accelerated motion are deduced and experimental verification discussed. Projectile motion is analyzed and the familiar parabolic law is obtained.

Much attention has been given by historians of science to the question of Galileo's understanding of the law of inertia. Nowhere does he clearly and precisely state the law, and it is often argued that he did not fully understand it. On the other hand, Drake[28] seems to be correct when he argues that in fact Galileo used the law of inertia as a matter of course in his analyses of motion and that the test should be not whether he gave a clear and unequivocal statement of the law, but rather whether he possessed an operational understanding of it, which he does seem to have had. As early as 1600 Galileo had written:

On a perfectly horizontal surface, a ball would remain indifferent and questioning between motion and rest, so that the least force would be sufficient to move it, just as any little resistance, even that of the surrounding air, would be capable of holding it still. From this we may take the following conclusion as an indubitable axiom: that heavy bodies, all external and accidental impediments being removed, can be moved in the horizontal plane by any minimal force.[29]

Experiment is never forgotten in *The Discourse,* but Galileo's preoccupation is clearly with the mathematical structure of the theory of mechanics, with experiment left largely with the limited role of verification. In his earlier work, Galileo much more strongly emphasized the role of experimentation, especially in criticism of the Jesuit failure to subject speculations about nature to the test of verification through sense experience. Galileo's later concentration on the hypothetical-deductive method in the *Two New Sciences* has led some commentators, especially Koyre, to characterize Galileo's scientific philosophy as Platonic, citing such famous statements as this one from *The Assayer:* "Philosophy is written in this grand book—I mean the universe—which stands continually open to our gaze, but it cannot be understood unless first one learns to comprehend the language and interpret the characters in which it is written. It is written in the language of mathematics, and its characters are triangles, circles, and other geometrical figures, without which it is humanly impossible to understand a single word of it: without these, one is wandering about in a dark labyrinth."[30] This argument, to the extent that it deals simply with labels to be attached to Galileo's methodology, is largely vacuous, but it does reflect the fact that in his later years Galileo's interests had turned strongly toward mathematical physics, and while he showed little interest in pure mathematics, he was grappling with problems of infinitesimals that were resolved only by the development of the calculus later in the century.[31]

Modern dynamics clearly begins with Galileo, as does mathematical physics. More than any man before him, Galileo understood the role of mathematics in describing the physical world, and during his life he formulated and fruitfully applied a very modern scientific methodology of abstraction from experiment, mathematical deduction and description, and empirical verification.

Conclusion

Galileo's achievements in astronomy were of great intrinsic value, but most of all they hastened the demise of scholastic Aristotelianism. The important of his work on the foundation of mechanics would be hard to overestimate, especially his experiments on and analysis of accelerated motion, his understanding that impetus is a *property* of a body in motion—not a *cause* of continued motion, and his studies of the relativity of motion.

In his work we see a growing understanding of force as the cause of a *change* in motion[32] and the emergence of a full understanding of the idea of instantaneous velocity. Yet his greatest achievement was the direction he gave to science, his modern views of careful experimentation, and his introduction of mathematical physics. He risked the wrath of authority while others were routed by the decree of 1616; he opposed his own authority to that of the Church Fathers on matters of natural philosophy and helped define science as an independent body of knowledge and methodology, free from the strictures of theology or moral philosophy. And yet he was a good Catholic and could never understand how elucidating God's eternal truths, by opening the eyes of men to the glories of God's creation, could be heresy.

CHAPTER 11

Newtonian Synthesis

Of Isaac Newton (fig. 11.1), Joseph Lagrange is supposed to have said, somewhat wistfully: "There could be only one Newton. There was only one world to discover."[1] Using the combination of observation and deductive reasoning that Galileo had applied to problems of terrestrial motion, Isaac Newton demonstrated mathematically what Galileo believed: that the physical universe is not two spheres but one world with one set of physical laws. Moreover, Newton believed that he was describing nature, not merely "saving the appearances." His laws or axioms of motion are elegant and inescapable, and together with universal gravitation explain planetary and terrestrial motions in a thoroughly convincing fashion. Within fifty years, and very nearly at the hand of one man,[2] physics became a rigorous and mathematically sophisticated science, the model for all the sciences—even the social sciences.

Whitehead has called the seventeenth century the "century of genius," for it is the century of Kepler, Galileo, Pascal, Descartes, Leibniz, Spinoza, and Newton. Kepler gave to science the laws of planetary motion that bear his name and conceived the first heliocentric planetary dynamics; Galileo shattered forever the distinction between the terrestrial and celestial worlds, formulated the first quantitatively correct laws of motion, and set the stage for a precise understanding of the idea of inertia. Descartes, the founder of modern philosophy, gave the first clear statement of rectilinear inertia, and with Galileo and Newton founded scientific rationalism. And yet Kepler buried his wisdom in neo–Pythagorean sun mysticism, Galileo remained to his death tied to circular planetary orbits and seems to have thought of inertia in terms of circular motion, while Descartes developed an a priori world system based on illusory vortices and rejected sense information in favor of reason as a way of understanding nature. How astonishing it is, then, that in 1687 Newton should have provided, in his *Mathematical Princi-*

Figure 11.1 Isaac Newton (1642–1727).

". . . the statue stood
Of Newton with his prism and his silent face,
The marble index of a mind for ever,
Voyaging through strange seas of Thought, alone."—William Wordsworth

ples of Natural Philosophy, so mature and rigorous a formulation of what
we now call classical mechanics, so convincing a description of the organization of the solar system, that for the next two centuries physics
busied itself with working out the consequences of this grand synthesis.
Newton produced what is indisputably the greatest work in the history
of science, a monumentally profound and totally convincing canoniza-
tion of the laws of mechanics and their application to the universe. It
was a grand synthesis of what had been learned and conjectured about
mechanics since Galileo, but it was far more than that. Remarkable as

the insights of Galileo are, the distance between these two seventeenth-century figures is enormous.

The astronomers had provided Newton with abundant raw material in the form of accurate observations of lunar and planetary motions, much of it distilled in the form of Kepler's laws. The mathematics of the day was inadequate for fully treating motion, as Galileo had found. So Newton invented the differential and integral calculus (which he called "direct and inverse fluxions"), a feat that alone was enough to indicate his genius. Newton was one of those men who, in Arthur Koestler's words, "as if they were charged with electricity, draw an original spark from any subject they touch, however remote from their proper field."[3] In addition to his accomplishments in mathematics and dynamics, Newton organized the new science of optics. His systematic investigations into the properties of light provided the basis for a century of optical research. Through his lifelong interest in alchemy, he became an expert in some aspects of metallurgy and in a primitive kind of chemistry.

In the broader cultural context the success of the revolution in physics encouraged the individualistic liberal philosophies that had been developing in Western Europe since the Renaissance. Newton became a symbol of the changes that enlightened rationalism could bring to the social structure.

Isaac Newton (1642–1727)

The drama of Newton's life is in ideas, not in events. His genius was recognized early, his acclaim as a scientist came in middle life, and he spent his declining years as the elder statesman of British science. Through it all he showed an unlikely combination of traits: fanatic dedication to scholarship, shyness and jealous pettiness, generosity, and a pronounced concern for social mobility. Because Newton came to be regarded as a paragon of English intellectual and personal virtues, candid biographies were not produced until late in the nineteenth century, and it is still not easy to distinguish fact from legend in the incidents of his life. Halley's view, although that of a friend, was not untypical: "Nearer the gods, no mortal may approach."

Isaac Newton was born on Christmas Day in 1642, the year of Galileo's death.[4] He grew up without a father, having an otherwise uneventful childhood in which he showed only moderate promise. Shy and introverted as a child, he developed into a suspicious, jealous man. He was not especially interested in his studies but excelled when challenged. Mechanical devices obsessed him, and he spent many hours in building models. He was somewhat frail, which led to self-absorption and meditation, growing up as he did sheltered in a lonely farmhouse. Newton attended a one-room school near the modest family estate until he was twelve years old. At that time he was sent to Grantham, six miles away, to prepare for the university. Three years later his mother called him home to help with the management of the farm. Lacking any competence for this work, he was soon back at Grantham. In 1661 he enrolled at Trinity College, Cambridge, where, under the tutelage of Isaac Barrow, first Lucasian professor of mathematics, he was soon making original discoveries in mathematics. Newton received his baccalaureate in 1665, his MA in 1668. Almost immediately after Newton's advanced degree was conferred, Barrow proposed that Newton assume the Lucasian chair, a position Newton held for the next 33 years.

In 1665, when he finished his undergraduate education, Newton withdrew to his mother's home to escape the plague. He spent more than a year there, during which period (his so-called *annus mirabilis*), according to his later recollection, he developed the methods of fluxions (the calculus), made his first steps toward understanding gravity,[5] and carried out experiments important for framing his corpuscular theory of optics. Newton's reputation as a mathematician grew steadily in the years that followed, and he was sought out by the amateurs of science who dominated the Royal Society. Yet despite his unmatched mathematical erudition and his dazzling cleverness in solving the problems that the members of the Royal Society posed for themselves, it was almost twenty years before Newton directed his efforts to a systematizing of his youthful discoveries in gravitation and motion. In 1685, after a series of sometimes acrimonious exchanges with Robert Hooke concerning the nature of the attraction between ponderous bodies,[6] and at the urging of Halley, Newton undertook to write down his thoughts on the dynamics. What emerged within two years was the monumental *Philosophiae Naturalis Principia Mathematica.* (*Mathematical Principles of Natural Philosophy*), which established physics as a mature science. The axiomatic approach and rigor of the *Principia* overwhelmed any lingering doubts about the validity of the Copernican system.

The years before the *Principia* were filled for Newton, not only with his academic duties and his researches into mathematics and optics, but also with alchemical experiments and with writing the earliest of his theological manuscripts.[7] His chemistry largely came to nothing; even Newton could bring little order to his observations without a quantitative atomic theory, which would not be developed for about 150 years. As to his theology, Newton wrote for his own edification and mental discipline. The manuscripts, which eventually totaled more than a million words, he mostly hid.

The effort of preparing the *Principia* told on Newton's health. In 1693 he suffered a nervous collapse, and though he recovered, he produced no really new ideas during the last thirty years of his life.[8] The *Opticks* was published during this period (1704), but it had been completed long before and appeared only after Hooke's death reduced the chance of controversy. Newton's views on light and optical phenomena came to be dominant within the new scientific community. The clarity and the accessible English of the *Opticks* had much to do with the broad acceptance, during most of the eighteenth century, of Newton's theory that light is corpuscular and not wave-like. Proponents of the wave theory had not only Newton's explanations but also his reputation to overcome.

Newton's latter years were centered on his public duties. His interest in the new party politics had brought him friendship of a prominent Whig, Charles Montague, later Lord Halifax. When Montague in 1696 secured for him the post of Warden of the Mint, Newton left Cambridge for London, apparently with no regrets. He never resumed the academic life but stayed in London as Master of the Mint, a highly paid sinecure, after spending three years as Warden supervising the historically important recoinage under King William III.

Newton in his London period acquired a coterie of scholars who championed the Newtonian philosophy and who prevailed upon him to publish his newer calculations as revisions of the *Principia*. Always a suspicious and close person even in his youth, and jealous of priority in his discoveries, Newton was not above using his supporters for his own purposes. The famous controversy with Leibniz[9] over priority in the calculus was aggravated by the partisan actions of Newton's young followers. The current view is that Leibniz and Newton each found the calculus independently and that Newton's understanding preceded Leibniz's by several years. Leibniz, however, certainly published first and developed the formalism further. Nevertheless, increasingly bitter exchanges led to accusations of plagiarism by both sides, and in 1712 Leibniz asked the Royal Society to resolve the question. Newton, who since 1703 had been president of the Royal Society, directed the investigations of a special committee, which solemnly found against Leibniz and for their president.

Newton remained vigorous and alert into his eighties, presiding with dignity at the Royal Society and working on revisions and additions to the *Principia*. Toward the end of his life he took a house in Kensington to avoid the London smog. He died there on March 20, 1727.

Newtonian Dynamics

Before examing the *Principia* to see how Newton developed his dynamics let us briefly review classical dynamics; we assume the reader
has had no prior contact with the subject. A logical starting point is the
crucial insight that motion can be understood only by concentrating on
changes in motion. The law of inertia ("Newton's First Law") states that
a body would continue to move in a straight line with constant speed
if nothing acted on it to change its motion. Stated positively, the first
law implies that a change in motion is caused by an agent. More rig-
orously, *changes in velocity* are caused by agents. Since velocity is specified by both speed and direction of motion, even circular motion at
constant speed is motion with changing velocity. Thus an agent is required to make the planets change direction but not to make them move.
This reverses the logic of Aristotelian cosmology. For Aristotle circularity in the heavens, motion itself, required a mover. The measure of
change in velocity is *acceleration;* [10] and Newton called the cause of motion "force." So the description of motion becomes "*acceleration* is caused
by a *force.*" Acceleration has both magnitude (a large acceleration is a
rapid change in velocity) and direction (acceleration is in the direction
of the *change* in velocity and so is not necessarily in the direction of the
velocity). The description of motion can be written as

$$\text{acceleration} \propto \text{force}$$
$$a \quad \propto \quad F$$

where the symbol $\propto$ means "is proportional to." In order to make this
proportionality into an equation (called "Newton's Second Law") we
multiply by a factor m, so that $ma = F$. The factor m, called mass, is a
measure of the resistance to change in motion. This can be seen by
writing the equation of motion as $a = F/m$. Hence a larger value for m
leads to a smaller acceleration. Since it is common experience that heavy
objects are harder to put into motion and harder to stop than light
objects are, mass must therefore be a function of the "amount of matter" in the object. Small masses can be directly compared against standard masses by using a simple two-pan balance or by studying their
vertical or horizontal accelerations when they are connected to a spring.
The simplicity of the property "mass" contributes crucially to the generality of the equation $F = ma$. Such qualities as color, shape, chemical
or physical state, do not influence the motion, except as the force may
depend on them.

From a known force and acceleration the mass can be determined.

But how is force to be defined—by the acceleration given to a particular mass? The argument is obviously circular. We now know that the apparently diverse "mechanical" forces—friction, collision, cohesion, etc.—are all fundamentally electrical. Newton had no way to know the importance of electricity in nature. Much of his mechanics dealt with mechanical or "contact" forces, but the triumph of his dynamics was in the quantitative formulation of *gravity*, the natural tendency of all objects to move each other. The fall of an apple (the famous story is mostly discounted now)[11] and the motion of the moon are each due solely to gravity, Newton proposed. But how is the *value* of the force of gravity to be assigned: which of the many properties of an object influences F in this case? Again, and marvelously, only the mass enters. Insofar as gravitational attraction is concerned, we can speak of objects as "masses," ignoring all other properties.

Newton also realized, as had Huygens and Hooke before him, that the gravitational attraction between two objects decreases with increasing separation. Consider two masses m and M, separated by a distance R (fig. 11.2). Newton's "universal" law of gravitation states that each mass experiences a gravitational force toward the other, of magnitude GmM/R^2 where G is a constant; the force depends only on the masses and their separation R, and not on any other property. In particular the force is quite independent of the state of motion of the objects.

Once the formula for the gravitational force is established, the motions of the objects can be found by solving the "equations of motion" that result when this formula is combined with Newton's Second Law. If two hypothetical masses are removed from each other so that their motions are entirely determined by gravitation, then for the mass m the acceleration is given by $ma = GmM/R^2$, or $a = GM/R^2$, toward M. Similarly, the acceleration of M toward m is given by $Ma = GMm/R^2$ or $a = Gm/R^2$. Notice that the equation of motion has the same form for each mass: $a = \text{constant}/R^2$.

It was the solution of this kind of equation that required the development of the calculus. The techniques for determining the subsequent

Figure 11.2 The gravitational forces that a pair of masses m and M exert on each other. The force that M exerts on m is equal to that exerted by m on M, but is opposite in direction. In either case, the magnitude of the force is GmM/R^2.

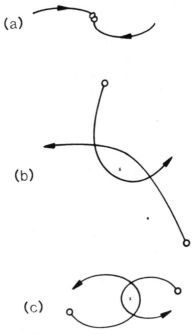

Figure 11.3 The three possible outcomes of the motions of two masses interacting through gravitational forces.

velocities and positions from a knowledge of the acceleration are not simple, but in the case of purely gravitational motion the possible motions are restricted. These are as follows: (a) the objects move toward each other until they touch (at which time nongravitational forces occur and the equation of motion changes form), or (b) they move past each other and continue to move apart, or (c) they "orbit" each other, neither touching nor escaping. These cases are illustrated in figure 11.3. In case (c) the orbit for each mass is an elliptical path about a point between them, the "center of gravity." The center of gravity is nearer the more massive object in direct proportion to its mass. If one of the objects is much more massive than the other, the orbit of the more massive object around the center of mass becomes much smaller than the orbit of the less massive object. The less massive object is then said to be orbiting its partner, and the center of gravity need not be mentioned. Our sun is much more massive than any of its planets, and so the influence of the planets on the sun's motion can be neglected. The motion of a moonless planet would be an ellipse about the sun, were it not for the influences of the other planets, which result in small but perceptible deviations from elliptical orbits for all the planets. A major moon, such as ours, causes larger deviations. The earth-moon motion can be thought of as compounded in the following way: the earth-moon center of gravity moves along an ellipse about the sun, while the earth and the moon circle their center of gravity.

The peculiar fact that the gravitational force is proportional to the mass of the body *acted on,* so that the acceleration depends only on the mass of the distant body, means that the effect of really distant masses (the stars) is scarcely noticeable, regardless of its magnitude (but see the discussion of Mach's Principle, chapter 14). Because the distance to the stars is essentially the same for the planet as it is for the sun, the stars affect the sun and the planets to the same degree, and there is no change in the planetary orbits, as long as the sun does not approach another star too closely. This picture already tells us a good deal about the motions of stars and galaxies. Typically stars "fall" past each other and the galaxies do the same thing; all the universe acts out a stately, eternal dance, punctuated by the rare fireworks of galactic collisions.

Since the universal gravitational force is proportional to the product of the two masses, the magnitude of the force is the same for each of the two interacting bodies. Newton was struck by this symmetry and proposed that all forces in nature occur between pairs of objects. This rule, called Newton's Third Law of Motion, is that "for every force on one object, there is another force, of the same magnitude but opposite direction, acting on another object."

If the Third Law is fully valid, and in a more general form it apparently is, then the dynamics can be extended to all of nature. The argument is as follows. No agent external to nature can exist, since by the Third Law everything that acts must be acted upon. But everything that responds to a force must have mass, according to Newton's Second Law. Hence nothing immaterial can act in the universe; Aristotle's "unmoved mover" cannot exist. This argument would not have bothered Newton, who preferred to think of God as transcendent lawgiver rather than as celestial carriage driver. The theological application of Newton's laws depends in any case on the assumption that the laws are exact and universal for all time. Newton maintained only that his laws served "pretty nearly"; for him the universe was less than 6,000 years old. During Newton's lifetime clergymen used the elegance and simplicity of Newton's laws to argue for a design in nature. In the eighteenth and nineteenth centuries, the Newtonian world as mechanism would be emphasized by the Cartesian philosophers.

Newton's Principia

No scientific work has had more influence on Western civilization than the *Philosophiae Naturalis Principia Mathematica,* published in 1687 and extensively revised in 1713 and 1726. Yet the book was and remains almost unreadable. There are two principal reasons for this inaccessibility, apart from the length of the book. First, Newton chose to organize the *Principia* in the manner of a mathematics treatise: propositions, lemmas, and corollaries follow each other in dizzying number, while the interdependence of the arguments make retrieval of a single result difficult. Secondly, Newton used none of the economy of the calculus in his derivations. The synthetic geometry he employed was intelligible to the trained mathematician of Newton's day, and Newton was comfortable with either method. Such a mass of geometry was indigestible, and it was not long until Newton's outline had been transliterated into the calculus. The course in mechanics thus defined has hardly been altered to the present day.

The *Principia* established a new method of persuasion: not really less verbal, but more quantitative and therefore less accessible to the general reader. The contrast with Galileo's *Dialogue* is arresting. By elegant and skillful argument, Galileo's protagonists literally persuade each other, and the reader, of nature's truth. In much of the *Principia* the numbers overpower the objections of dissenters, whatever the merits of the prose. We can perhaps get some feeling for the power of Newton's mathematical philosophy by a gingerly examination of the organization and emphasis of the *Principia.*

Newton begins by defining the quantities that enter his description of motion: mass, force, momentum, acceleration. He then devotes several pages to a discussion of his ideas about space and time. He emphasizes the difference of his "absolute" space from the Aristotelian "place" in a famous passage.

Absolute space in its own nature, without relation to anything external, remains always similar and immovable. Relative space is some movable dimension or measure of the absolute spaces; which our senses determine by its position to bodies; and which is commonly taken for immovable space; such is the dimension of a subterraneous, an aerial, or celestial space, determined by its position in respect of the earth.[12]

The question of relative and absolute motion that arises from the introduction of absolute space is discussed next. Newton attaches a profound philosophical importance to absolute motion, but requires that when absolute and relative motions cannot be distinguished the distinction can be dropped "without any inconvenience in common affairs." Newton gives examples of situations in which a system apparently at rest can be determined, from the forces acting on it, to be actually rotating. The broader questions of relative motion are not raised again, despite the promise "to obtain the true motions from their causes," which ends this introductory section. Not for two hundred years were these questions to be resolved, and then only by Albert Einstein.

The preliminaries are completed in a brief section in which the three "laws of motion" are stated succinctly, as if geometric axioms, along with the rules for combining vector quantities. The bulk of the *Principia* (we refer to the third and most comprehensive edition) is divided into three parts. In book I, "The Motion of Bodies," Newton derived the motions of a particle subject to various kinds of force. The inverse square attractive force is of course given special consideration, since the gravitational force takes this form. The paths produced by the forces examined include conic sections and other, more complex curves. The derivations of the properties of these curves are inserted wherever they are needed throughout book I. The effect of this procedure may be judged from the reproduction (fig. 11.4) of Newton's derivation of Kepler's first law (elliptical orbits) from the inverse square force law.

Whether Newton had, as he said late in life, speculated about the gravitational pull of the earth as early as 1666, he did show in the *Principia* that the effect of the earth, treated as a sphere, is the same as if its mass were concentrated at its center. This simplification is crucial for the application of the inverse square law to motion of objects at and near the earth. In proving this and related results for extended bodies Newton explicitly drew upon the logic, if not the notation, of the calculus. Angles are required to vanish so as to provide simple limiting ratios, and solid spheres are subdivided into "innumerable spherical shells" and the effects of the shells added.

Book II, "The Motion of Bodies in Resisting Media," is somewhat less formal than book I. Newton wanted to show that, for terrestrial objects, such as projectiles and the pendulum, the deviations from the simple motions obtained from them in book I could be entirely and quantitatively accounted for by the resistance of the air. Newton also describes various measurements he made to test air resistance. These descriptions, too closely knit to be excerpted here, document his remarkable aptitude for experimentation. After treating the motions of solid bodies through a fluid medium, Newton considers the dynamics of the fluid itself, treated as continuously distributed masses. These results—for example, the motion of the part of a fluid near a rotating

SECTION III

The motion of bodies in eccentric conic sections.

PROPOSITION XI. PROBLEM VI

If a body revolves in an ellipse; it is required to find the law of the centripetal force tending to the focus of the ellipse.

Let S be the focus of the ellipse. Draw SP cutting the diameter DK of the ellipse in E, and the ordinate Q*v* in *x*; and complete the parallelogram Q*x*PR. It is evident that EP is equal to the greater semiaxis AC: for drawing HI from the other focus H of the ellipse parallel to EC, because CS, CH are equal, ES, EI will be also equal; so that EP is the half-sum of PS, PI,

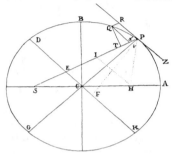

that is (because of the parallels HI, PR, and the equal angles IPR, HPZ), of PS, PH, which taken together are equal to the whole axis 2AC. Draw QT perpendicular to SP, and putting L for the principal latus rectum of the ellipse (or for $\frac{2\,BC^2}{AC}$), we shall have

$$L \cdot QR : L \cdot P\textit{v} = QR : P\textit{v} = PE : PC = AC : PC,$$

also, $L \cdot P\textit{v} : G\textit{v} \cdot P\textit{v} = L : G\textit{v}$, and, $G\textit{v} \cdot P\textit{v} : Q\textit{v}^2 = PC^2 : CD^2$.
By Cor. II, Lem. VII, when the points P and Q coincide, $Q\textit{v}^2 = Q x^2$, and $Q x^2$ or $Q\textit{v}^2 : QT^2 = EP^2 : PF^2 = CA^2 : PF^2$, and (by Lem. XII) $= CD^2 : CB^2$. Multiplying together corresponding terms of the four proportions, and simplifying, we shall have

$$L \cdot QR : QT^2 = AC \cdot L \cdot PC^2 \cdot CD^2 : PC \cdot G\textit{v} \cdot CD^2 \cdot CB^2 = 2PC : G\textit{v},$$

since $AC \cdot L = 2BC^2$. But the points Q and P coinciding, 2PC and G*v* are equal. And therefore the quantities $L \cdot QR$ and QT^2, proportional to these, will be also equal. Let those equals be multiplied by $\frac{SP^2}{QR}$, and $L \cdot SP^2$ will become equal to $\frac{SP^2 \cdot QT^2}{QR}$. And therefore (by Cor. I and V, Prop. VI) the centripetal force is inversely as $L \cdot SP^2$, that is, inversely as the square of the distance SP. Q.E.I.

The same otherwise.

Since the force tending to the centre of the ellipse, by which the body P may revolve in that ellipse, is (by Cor. I, Prop. X) as the distance CP of the body from the centre C of the ellipse, let CE be drawn parallel to the tangent PR of the ellipse; and the force by which the same body P may revolve about any other point S of the ellipse, if CE and PS intersect in E, will be as $\frac{PE^3}{SP^2}$ (by Cor. III, Prop. VII); that is, if the point S is the focus of the ellipse, and therefore PE be given as SP^2 reciprocally. Q.E.I.

With the same brevity with which we reduced the fifth Problem to the parabola, and hyperbola, we might do the like here; but because of the dignity of the Problem and its use in what follows, I shall confirm the other cases by particular demonstrations.

Figure 11.4 Newton's derivation of Kepler's first law. Newton, *Mathematical Principles of Natural Philosophy*, A. Motte, trans., revised by F. Cajori (Berkeley: University of California Press, 1934, reprinted 1962).

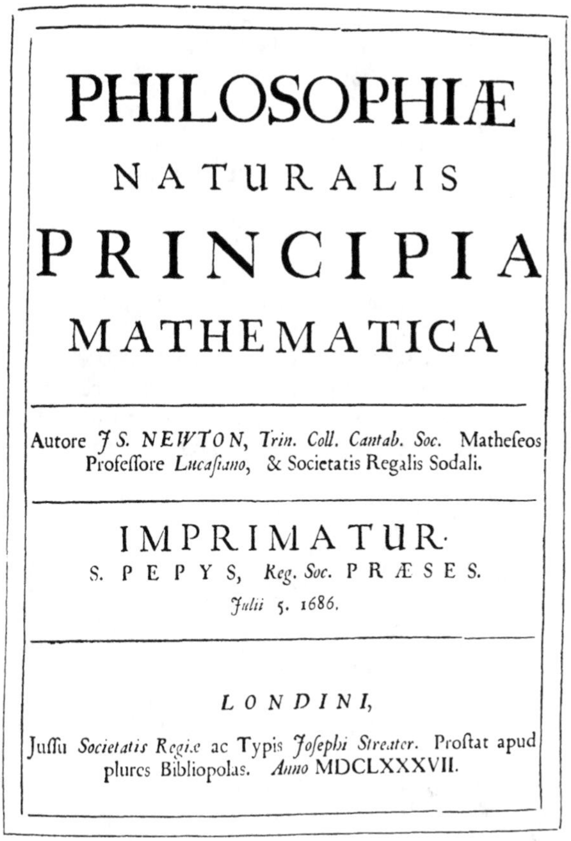

PHILOSOPHIÆ

N A T U R A L I S

P R I N C I P I A

MATHEMATICA

Autore *J S. NEWTON*, *Trin. Coll. Cantab. Soc.* Mathefeos
Profeffore *Lucafiano*, & Societatis Regalis Sodali.

IMPRIMATUR·
S. P E P Y S, *Reg. Soc.* P R Æ S E S.
Julii 5. *1686.*

L O N D I N I,

Juffu *Societatis Regiæ* ac Typis *Jofephi Streater.* Proftat apud
plures Bibliopolas. *Anno* MDCLXXXVII.

Figure 11.5 Title page of the first edition of the *Principia*, 1687. Source: same as fig. 11.4.

cylinder, defined hydrodynamics. Many of the results in hydrodynamics are of interest in their own right, but in addition there are compelling historical reasons for Newton's interest in fluids. One of of these is the evident importance of hydrodynamics in Newton's corpuscular optics—and to the wave theory of sound, which Newton rather casually developed in a few pages. But a second and more fundamental reason for treating hydrodynamics was the need to refute Descartes's theory of vortices. Newton is at pains to demonstrate that a material vortex could not drive the planets in their observed motions. After the properties of vortices are demonstrated in a series of propositions on the circular motions of fluids, Newton concludes that Descartes's theory is inconsistent with Kepler's third law, that the periods of the planets increase as the 3/2 power of their mean distance from the sun. The argument is a felicitous example of the methods of the new philosophy. Advocates of any non-Newtonian cosmology must account for the 3/2th power through mathematical proof or abandon the field. For the modern reader, vortices are polished off neatly enough by Newton's proof, complete with characteristically overstated numerical precision, that the atmosphere is thin, so that the space among the planets is empty of any matter from which a vortex could be formed.

It is shown in the Scholium of Prop. XXII, Book II, that at the height of 100 miles above the earth the air is more rare than it is at the surface of the earth in the ratio of 30 to 0.0000000000003998, or as 75000000000000 to 0.1, nearly. . . . And therefore the celestial regions being perfectly void of air and exhalations, the planets and comets meeting no sensible resistance in those spaces will continue their motions through them for an immense tract of time.[13]

In the first two books Newton does not treat the motions of the planets per se but lays the groundwork by considering generally the motion of "bodies." Let Newton introduce the theme of book III:

In the preceding books I have laid down the principles of philosophy; principles not philosophical but mathematical: such, namely, as we may build our reasonings upon in philosophical inquiries. These principles are the laws and conditions of certain motions, and powers or forces, which chiefly have respect to philosophy; but, lest they should have appeared of themselves dry and barren, I have illustrated them here and there with some philosophical scholium, giving an account of such things as are of more general nature, and which philosophy seems chiefly to be founded on. . . . It remains that, from the same principles, I now demonstrate the frame of the System of the World.[14]

One after the other the astronomical phenomena are succinctly matched against the predictions of book I. The planetary laws of Kepler need only the briefest exposition and the proof of Copernican system and the inverse square law begins, "The former part of the Proposition is manifest from Phen. V. and Prop. II, Book I; the latter from Phen. IV and Cor VI, Prop IV, of the same book. . . ."[15] The tides, the moons of Jupiter, the precession of the equinoxes are disposed of in much the same way. The almost liturgical quality of book III is interrupted by two long calculations, one on comets and the other a treatment of the details of the moon's motion, which were late additions. The calculation on comets, which used observational data to prove that comets follow gravitational orbits, finally removed these dramatic objects from the realm of the supernatural. The lunar calculation set the pattern for future calculations in celestial mechanics and, incidentally, occasioned years of bitterness between Newton and John Flamsteed, the Astronomer Royal, whose measurements Newton used without giving adequate credit. The smallest corrections to the orbits of the planets, except for Mercury, were eventually found to conform to the Newtonian rules. But the job took more than a century to complete and involved the discovery of three new, distant planets.

The *Principia* ends with the famous "General Scholium," which Newton appended to the second edition, in 1713. The General Scholium has been the source of much controversy, for in it Newton discusses the origin of gravity. Though his opinion is that, "This most beautiful system of the sun, planets, and comets, could only proceed from the counsel and dominion of an intelligent and powerful Being," whom he names as the "Lord God or Universal Ruler" (adding in the third edition, published when Newton was 84, "the God of Israel, God of Gods, Lord of Lords"), nevertheless Newton finally refuses to assign a first cause for gravity.

Hitherto we have explained the phenomena of the heavens and of our sea by the power of gravity, but have not yet assigned the cause of this power. This is certain, that it must proceed from a cause that penetrates to the very centres of the sun and planets, without suffering the least dimunution of its force. . . . But hitherto I have not been able to discover the cause of those properties of gravity from phenomena, and I frame no hypotheses, for whatever is not deduced from the phenomena is to be called an hypothesis; and hypotheses, whether metaphysical or physical, whether of occult qualities or mechanical, have no place in experimental philosophy.[16]

Newton's refusal to "frame hypotheses," by which he meant to propose untestable theories, is restricted to the question at hand. His definitions of space and time are clearly hypotheses, since they are not verifiable by his own methods. And one proposition in book III is even labelled "hypothesis." But the disclaimer was interpreted broadly by the mathematical physicists who came after Newton. If book III of the *Principia* is on the whole as full of resolution as the last movement of a great symphony, then the last paragraphs of the General Scholium could be linked to the lingering final chord, whose dying away is a reminder of the finitude of all man's works. For Newton concludes the *Principia* with a catalogue of phenomena "that cannot be explained in a few words."

And now we might add something concerning a certain most subtle spirit which pervades and lies hid in all gross bodies; by the force and action of which spirit the particles of bodies attract one another at near distances, and cohere, if contiguous; and electric bodies operate to greater distances, as well repelling as attracting the neighboring corpuscles; and light is emitted, reflected, refracted, inflected and heats bodies; and all sensation is excited, and the members of animal bodies move at the command of the will, namely, by the vibrations of this spirit, mutually propagated along the solid filaments of the nerves, from the outward organs of sense to the brain, and from the brain into the muscles. But these are things that cannot be explained in a few words, nor are we furnished an accurate determination and demonstration of the laws by which this electric and elastic spirit operates.[17]

Muscle and nerve, the organs of sense, the brain—the explanations of these systems pose for physical science problems that may never fully be solved. But it would be historically inaccurate to assume that Newton was suggesting the extension of the dynamical calculations to such living things. Rather, he was here alluding to a most fundamental question: How are forces transmitted from one body to another? Do distant objects simply act directly on each other, or does some medium intervene? Newton asserts that there is something that serves to join objects; it is "a certain subtle spirit which pervades . . . all gross bodies." This "electric and elastic spirit," which was called "ether" (though its properties were only superficially related to Aristotle's planet-moving ether) would transmit all forces, the mysterious ones discribed in the paragraph above as well as the simple, universal gravitational force. The debate over action at a distance versus the ether continued into the twentieth century and provided the context for Einstein's revolutionary rethinking of time and space.

Scientific Rationalism: Newton and the Enlightenment

Newton's approach to natural philosophy was typical of educated Protestants of his day, in that he regarded rational inquiry as a valuable tool for discovering the Divine plan of the world. Such intellectual humility came to be a difficult posture for the scientific philosophers. For Galileo's generation pride of achievement was a fault, confessed only under duress. After Newton there was no need for modesty. Confidence in scientific rationalism reached a peak in the eighteenth century, the period known as the Enlightenment. Few Enlightenment philosophers fully understood the mechanics, however (Voltaire being something of an exception), and more often the method was invoked rather than followed. John Locke is an example. In politics and other science of man, new "laws" of human behavior were discovered. Their impact on Thomas Paine, and their consequent importance in the American and French revolutions, is well known. But the social scientist gave a different meaning to "natural laws" than had Newton, who used the term to characterize what is, not what ought to be. As Gillispie has pointed out, Newton would have been a lawgiver in the sense of Enlightenment social science if he had established the law of gravity "by persuading the planets to obey the inverse square relationship in their own interest."

Newton, the consummate scientist, validated the hopes of the scientific rationalists without accepting their philosophies. Francis Bacon (1561–1626), though not a scientist, had been the prophet of the new science. Bacon advocated a version of the "scientific method," that natural laws can be discovered by gathering and systematizing a sufficient number of facts. Bacon's greater emphasis on the empirical and the experimental rather than on the theoretical was not to the taste of either Galileo or Newton, both of whom preferred to have their ideas in hand before undertaking an experiment. For lesser men and for less mature sciences such as eighteenth-century chemistry, the facts more often than not preceded the explanation, just as Bacon had said. In his reasons for championing science Bacon was even more prophetic. He believed that organized, state-subsidized science could bring the subjugation and control of nature. This vision of technological progress did not begin to be realized until the nineteenth century. Technology and science are so linked in the twentieth century that it requires a conscious effort to recall that Newton's career produced nothing practical.[18]

Rene Descartes (1596–1650), called the father of modern philosophy, had been unwilling to wait until scientific progress should produce a fully understandable world system. Descartes invented a cosmology that was influential enough to be included in university curricula for fifty years or longer before it was supplanted by Newtonian dynamics. Descartes argued from geometry to a mechanistic universe. Imagining real space as marked by the intersecting straight lines of the rectangular Cartesian coordinates he had invented, he was led to what Newton later named his first law (straight-line motion for undisturbed bodies) from Euclidian geometry (Parallels meet only at infinity). Descartes's dynamics was simple: every motion, and therefore all structure, was produced by successive collisions within a plenum. Regardless of its initial motion, matter evolved under continued collisions to a vortex with the heaviest matter (the sun) at the center. Further out in the vortex the planets were carried along by a transparent but material medium. The Cartesian physics is completed by the assumption of a constraint, namely, that the total amount of motion in the universe cannot be changed. The scheme was entirely qualitative; no equations of motion were proposed. This cosmology wedded Democritus and Aristotle to Copernicus. Yet even his continental contemporaries saw his science as "purely imaginary physics, a philosophical romance." [19]

Like Aristotle, Descartes made his physics depend on his metaphysics; his whole philosophy assumed a mechanical world. In his treatment of the mind–body problem, for example, the soul cannot change the amount of motion, but it can redirect motion. In this way the soul is limited to actions consistent with the Cartesian physics. An inescapable implication of the Cartesian dynamics is an infinitely extended universe. Since only a collision could turn back a particle, there could be no boundary to the world. Descartes apparently accepted this heretical inference, though he found it expedient not to publish it. For Descartes the world was a "cosmic hydraulic system," "as full of itself as an infinite egg." Newton accepted neither the vortex nor a fully mechanistic universe. He was able in the *Principia* to demolish Cartesian physics, but the Cartesian clockwork universe survived in the "Newtonian world view."

The successes of the Newtonian method encouraged the Cartesian school to an even more rigidly mechanical world view. Previously their mechanistic biology had extended to animals but not to man. Lacking a soul, an animal could not experience pleasure or pain, and so could be petted or tortured with indifference. The soul described by Descartes is an unmoved mover, and so would violate Newton's third law; the Cartesian philosopher Geulinex therefore replaced the free soul with the "two clocks." He proposed that the soul (or mind) is like an independent clock synchronized with the clock of the material world, a soul programmed to feel pain at the same time the body is injured, to see only when the body opens its eyes, not interacting with the body. If this theory is an illustration of the lengths to which some eighteenth-century thinkers were driven in conforming their philosophies to the new dynamics others, such as the Enlightenment optimists, had held their premises despite the restrictions of the dynamics. While Plato and Aristotle had found the deterministic world proposed by Democritus unthinkably regular, for Leibniz and Baruch Spinoza, determinism implied not oppression but security. If events could not possibly be other than they are, they argued, and since God is good, then the universe must be organized on a principle of minimized evil: this is the best of possible worlds.

The brief examples we have given perhaps suggest how the influence of the new dynamics began rapidly to diffuse throughout society, and not only in philosophy. In literature the impact of the scientific revolution was dramatic, although the direct effect of Newton is not so easily separable, and in the arts the Newtonian influence was more subtle yet. Despite Newton's aura, the philosophers increasingly found little common ground for discussion with the new physicists. The age of specialization had begun. From the eighteenth century onward no one person was expected to grapple with the whole of existence. In the new sciences modeled on empiricism and the Newtonian method—chemistry, biology, geology, psychology—the best thought was that of brilliant specialists. There could never again be an Aristotle.

The Newtonian dynamics was applied with great success to celestial mechanics, the problem of the planets. Many of the greatest mathematicians of the eighteenth century devoted much effort to the development of techniques of celestial mechanics and in actually analyzing data— Euler, Lagrange, and Laplace in particular. Laplace's *Treatise on Celestial Mechanics,* published in five volumes between 1799 and 1825, is one of the classic works on the subject. Bessel, Jacobi, and somewhat later Poincaré, also contributed to the techniques of perturbational calculation of planetary orbits, and as early as the end of the eighteenth century, astronomers could boast an almost complete understanding of the most detailed planetary motions. The precision of perturbational calculations of the planetary orbits came to be so great that by mid-nineteenth century the existence of a new planet could be predicted on the basis of such calculations.

Newton never expected that his laws would be used to follow the motions of the planets far back into time. The continental mathematicians had no such prejudice and saw that their successes in celestial mechanics made plausible a Newtonian cosmogony in which the universe had evolved to its present form as a natural result of gravitation, a process that would require an immense time. One of the first men to advocate the extreme age of the earth was the great German philosopher Immanuel Kant. In 1755 Kant published his *Theory of the Heavens,* in which he maintained that the planets had coalesced out of a primeval cloud of gas. These speculations were, however, not supported by quantitative arguments. Later in the same century Laplace proposed and defended a similar mechanism for the birth of the solar system. Increasingly cosmology came to be the province of astronomy and physics. Newton had demonstrated the frame of the system of the world better than he knew, though much of his language would be supplanted in the twentieth century.

CHAPTER 12

Widening Horizons: Einstein

Maturity and Crisis

The richness and variety of the discoveries in physics in the two centuries after the publication of the *Principia* in 1687 and the prolific successes of observational astronomy during the same period were extraordinary. Some understanding of these achievements in physics is necessary in order to see where the twentieth-century revolutions in physics stem from. We shall turn again to the heavens in subsequent chapters, but for the moment we begin our development of the modern view of the universe in the most direct way, through the life and works of Albert Einstein, who was born in Ulm, Germany, in 1879 and died at Princeton in 1955.

Who was Einstein, that he should have been the gatekeeper to the twentieth century, the one person most instrumental in altering Newtonian physics for atoms and for stars, who shattered conventional ideas about space and time? Born in Germany of bourgeois Jewish parents, he was educated in Catholic and public schools there and, his family having moved to Italy, at the Zurich Polytechnic (after a year of schooling occasioned by his failing the entrance exam at the Poly). He early showed the nonconformity that was to be a persistent trait, so that he easily gained competence in the mathematical subjects he liked but often failed to impress his professors.[1] Although he had thought to become an engineer, at Zurich he was drawn to the more modern topics of thermodynamics and electrodynamics, the grand achievements of nineteenth-century physics. Soon after completing his degree he won recognition within the scientific community by a series of brilliant and controversial papers built on these two subjects and written at nights and on Sundays, the only time available to him from his job as a patent examiner in Bern. Within fifteen years after leaving Bern in 1909 for his first academic position, he was a Nobel Laureate and a world celebrity, and the stereotypic image of him as the kindly, idealistic, but remote professor had begun to take hold among the public.

There is drama in Einstein's later life, in the emigration to the United States in 1933, in his involvement with antimilitarist and humanitarian causes even while the new physics was being used by governments in an effort to produce the nuclear bomb, in his being offered the first presidency of the infant state of Israel. But our interest is in the ideas that drew Einstein's attention from his sixteenth year, ideas that grew into the various realizations of "relativity theory." It is here that the mathematical basis for modern cosmology is to be found.

The year of Einstein's birth, 1879, is a convenient date for examining the successes and failures of the Newtonian physics in its mature form. The discipline had grown to encompass an extraordinary range of phenomena. In addition to the triumphs in celestial mechanics described in the previous chapter and laboratory-scale results in particle and fluid motions and in acoustics, there were the beginnings of systematic engineering applications, especially those related to the design of structures and machines. Enriched by the introduction of the energy concept and the principle of conservation of energy, the dynamics had been extended to include the study of temperature-dependent processes: heat transfer, melting and freezing, pressure effects, and the like. Out of the great variety of such measurements there emerged a new and elegant branch of mathematical physics—thermodynamics—whose equations express the most general properties of matter based on only two or three experimental results, introduced at the outset as axioms. This generality and independence of a large mass of empirical data has allowed thermodynamics to survive to the present day, augmented but unscathed.

It was generally accepted that thermodynamic effects must be grounded in the behavior and structure of the atoms that comprise matter, but no satisfactory Newtonian theory of atomic structure was forthcoming. Nor was one likely to arise from thermodynamics, whose variables are determined by averages over enormous numbers of atoms, so that detailed atomic properties are masked. A tantalizing feint at a microscopic theory of matter had been part of the new electrodynamics, the unification of electrical and magnetic effects with optics. Here there was hope that the structure of matter could be found by tracing the source of electrodynamic effects to the atomic scale, to the carriers of electrical charge.

The leading British electrodynamicists, Michael Faraday (1791–1867) and James Clerk Maxwell (1831–1879), were committed to the field concept. A "field" in this context is the region around a source of electromagnetic force (or, by extension, a source of any kind of force, gravitation, for example).[2] The idea is that forces are caused, not by direct action between two objects, but when an object enters the region of influence (the field) of another. The field thus intervenes and makes the space between the objects (and within them) an active part of the dynamic process. When Newton ended the *Principia* with a reference to an all-pervasive "electric and elastic spirit," he was casting his vote, implicitly, for the field concept against action at a distance.[3] But he wisely saw that the choice cannot be made from studying ordinary gravitational situations: the earth's gravitational field is already present, is static, and the mass distribution changes quite slowly. The electrodynamicists introduced electric and magnetic fields as mediators of the effects of intangible, time-dependent sources. They made great progress, both experimentally and theoretically, and they considered the fields to be "real," as indeed anyone might who has played with even a weak magnet.

In Faraday's view the field was the *only* reality, and solid particles were dispensable. Atoms were simply places where the field was especially dense, and thus the sources should be described by the same equations as predict the forces. Maxwell obtained differential equations satisfied by the fields that are in widespread use today. But his attempts at developing a detailed electromagnetic substructure of matter, in terms of "vortices" and "tubes" of electromagnetic influence, were unsatisfactory and were soon abandoned.

Stripped of any detailed imagery of underlying sources, Maxwell's electrodynamic field equations are comparable in elegance to the equations of thermodynamics, in that each theory consists of precise mathematical relations whose validity is independent of the details of microscopic structure. Many physicists of the late nineteenth century (including certainly the experimenters who investigated the predictions of Maxwell's equations) were uncomfortable with this unexpected austerity. J. L. Synge has described the unadorned equations of Maxwell as substituting for "the robust body of the Cheshire cat . . . only a sort of mathematical grin."[4] Maxwell's vortices had been visualized as mechanical disturbances in a continuous elastic structure identified with the ether. During the last decades of the nineteenth century a substantial effort was made to relate electromagnetic effects to the ether. Especially for light, it seemed that some kind of supporting medium was essential. Some of the most able of physicists, H. A. Lorentz of Leiden and the American Albert Michelson among them, were drawn to electrodynamics, and in the first years of the new century the debate about Maxwellian physics was intense.

When the young Einstein, who had begun reading electromagnetism textbooks while still a student in Zurich, took his first regular job, in 1901 at the Swiss Patent Office in Bern, he was remote from the give and take of the controversy but was free to pursue his interests in mathematical physics and to hone his arguments in long discussions with a few like-minded friends. Although his isolation from academic associations had not been by choice, Einstein found the experience at Bern salutary: "I soon learned to sort out that which is able to lead to fundamentals and to turn aside from everything else, from the multitudes of things which clutter up the mind and divert it from the essential."[5] Soon he was producing papers that broke new ground in thermodynamics. One of these earned him his doctorate from the University of Zurich, while another, published in 1905, yielded in a most surprising way the quantum hypothesis (and eventually earned him the Nobel prize).[6] His first paper on electrodynamics was published in the same year. Entitled "On the electrodynamics of moving bodies," it embodied the special theory of relativity, a new description of space and time.[7]

From the theory of relativity came a fundamental revision of the laws of dynamics, and it demanded a radically new understanding of space and time. Furthermore, its geometrical underpinning pointed in the direction of a new theory of gravity and its emphasis on invariance principles ("Lorentz invariance") encouraged the search for more general symmetries. In these ways, the special theory contributed crucially to our understanding of the universe. It was a perfect vehicle in terms of which Einstein could work out his predilection for symmetry and beauty in physical theory. He published many papers on other topics, but he gave to the questions raised by the relativity theory his best thought for forty years.

Special Relativity

Although the special (or restricted) theory of relativity is surprisingly free of mathematical complexity, it has an unassailable logic to it. Only in his later forays into "unified field theories" was Einstein forced to incorporate truly abstruse mathematical ideas. We can thus summarize his 1905 paper, pausing only to comment that of the many popular treatments of relativity, none are clearer or more charming than his own.[8]

Einstein began with a pair of postulates, from which the far-reaching consequences of special relativity are deduced. Observation or experiment is left to decide, if possible, the validity of the theory. This approach, so characteristic of Einstein's early papers, is but one path to scientific discovery.

Einstein stated two general assumptions (postulates):

Principle 1: The physical laws have the same form for all inertial observers.

Principle 2: The measured speed of light is the same for all inertial observers, independently of the motion of the source or observer.

The first of these is the "principle of relativity," made "special" by its restriction to inertial observers, that is, those for whom Newton's first law holds. An observer on the earth is not inertial, since as seen from the earth the stars go in circles without any force acting on them. Another way to define an inertial observer is to say "nonaccelerated observer." Each inertial observer, then, is moving at a constant speed relative to all other inertial observers. The principle of relativity, a deceptively "reasonable" if very fundamental conjecture about the symmetry of the world, had been discussed earlier by Henri Poincaré (although Einstein was apparently unaware of this). The second postulate, the universal constancy of the speed of light, was decidedly radical, because it conflicted not only with commonsense notions of wave propagation but also with ether theories of electrodynamics. If light is wave motion, its velocity should change when one is moving toward the source through the medium of propagation or moving away from the source. The paradoxical behavior asserted by the second postulate can be exchanged, as Einstein showed, for the statement that there is no universal or absolute time.

In order to demonstrate the logic of special relativity, Einstein first examined the concept of simultaneity. If two events occur at different places, judging whether they are simultaneous requires synchronized clocks. In special relativity, clocks can readily be synchronized with light signals, since by the second postulate, the speed of light is constant. Two clocks at rest with respect to each other can be synchronized by sending a light pulse from the first clock to the second and correcting for the signal time: distance/c. This is all straightforward enough if the clocks are at rest. Suppose, however, that the clocks are in motion with respect to another coordinate system ("the laboratory") with its own set of clocks.

We consider a pair of clocks attached to the ends of a rigid rod, moving to the right through the laboratory, as shown in figure 12.1. A set of "stationary" clocks, fixed in the laboratory, synchronized by the

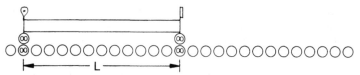

Figure 12.1A   Relativity of Simultaneity. In this paraphrase of Einstein's first published thought experiment on relativity, the constancy of the speed of light is used to demonstrate that simultaneity is not an absolute, but depends on the state of motion of the observer. The rod and its two clocks are moving to the right at one-half the speed of light ($v = 0.5c$). The laboratory clocks are synchronized and the moving clocks keep "laboratory time," in that each reads the same time as the nearest laboratory clock. A light flash leaves the left end of the rod at a laboratory time of 12:00.

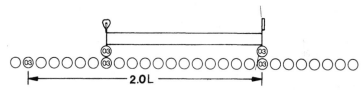

Figure 12.1B   The light flash overtakes the right end of the rod at 12:03, after traveling a lab distance of 1.5 times the length of the rod. A mirror reflects the light back along the rod.

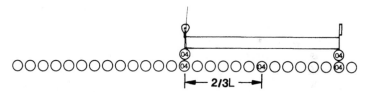

Figure 12.1C   Since the left end of the rod advances one-half the rod's length during the return trip, the lab distance traveled by the light is one-third that for the first segment, and the flash reaches the left end of the rod at 12:04 laboratory time. The clocks on the rod, and therefore all the lab clocks, fail the test of synchronization for observers on the rod, who divide the light's trip into equal halves (equal distances traveled in equal times), and so would require the light flash to reach the right end at 12:02 for synchronization. Time intervals depend on motion relative to clocks.

method just described, is arrayed along the path of the rod. Further, the moving clocks are "slaved" to the laboratory clocks they move past, so that each reads the same as the laboratory clock it is adjacent to. It is not hard to show that in this situation the moving clocks, synchronized for laboratory observers, are not synchronized for observers moving with the rod. If the synchronized laboratory clocks are used to initiate a pair of simultaneous events (simultaneous in the laboratory), such as flashes of light, then because the moving clocks are found not to be synchronized, the events will not be seen to be simultaneous to an observer moving with the rod.[9]

This result is called the "relativity of simultaneity." If two events at different places are simultaneous for stationary observers, they will not be for any uniformly moving observer. Clearly, in a world where the speed of light is a constant, time intervals are not absolute but depend on the motion of the observers. Inertial observers disagree about time intervals between events unless they are at rest with each other. Extending this reasoning, Einstein showed that if two observers are in relative motion, each finds two clock rates: his own local or "proper" time and that of the system moving with respect to him. The moving clock is found to be slowed by a factor $(1/\sqrt{1-v^2/c^2})$, which is imperceptibly different from unity for low relative speeds but which becomes large near the velocity of light. This effect is called "time dilation" and is by now a commonplace in modern physics. In high-energy physics laboratories, where the decay time for unstable particles may be slowed a hundredfold by time dilation, experimental areas are designed to accommodate such relativistic effects (see appendix).

A similar modification of spatial intervals or lengths as measured by a pair of observers in relative motion occurs. The length measurement can be thought to consist of marking the separation of the two ends of a moving rod at the *same time*. The result will be smaller than the rest or "proper" length by the same factor that appears in time dilation. This explains, by the way, how the decaying unstable particle reaches the final target area without decaying, as seen in *its own reference frame*. To an observer moving with the particle, the distance from where it was produced to the target area is contracted by precisely the same factor by which the time interval was dilated as seen by an observer in the laboratory. This "length contraction" effect is, of course, also significant only at speeds near c.

A further unexpected implication of the constancy of the speed of light arises in the measurement of relative velocities. Our usual experience is that, when we move away from a disturbance, it will overtake us more slowly than if we were stationary. This is true, for example, with sound. Not so, however, with light. No matter how fast an object recedes from a source of light, the light will overtake it at velocity c. Careful examination of this dilemma leads to the conclusion that speeds greater than c are forbidden by the relativity postulates.[10] The speed of light is a limiting velocity rather than simply the propagation velocity for some particular form of energy.

Spatial and temporal intervals and relative velocities are altered in special relativity as compared to their commonsense usages. Carefully prescribed procedures must be given for the measurement of these quantities, which include specifying the coordinate system in which the measurements are being made. Similar modifications of dynamics are required by special relativity. For example, Einstein showed in the first relativity paper that if "mass" is to mean the factor that multiplies acceleration in Newton's second law, then it must depend on velocity. The observation of this effect, the relativistic increase in mass for electrons, was the first direct laboratory confirmation of special relativity.[11] A similar analysis of energy in special relativity resulted in two papers, the second of which proposed the equivalence of mass and energy. The equivalence takes the form $E = mc^2$, which results in a generalization of energy conservation to "mass-energy conservation": mass-energy cannot be destroyed in any process; it can only be changed in form. All forms of energy, including electromagnetic waves and the thermal motion of atoms, contribute to mass, or inertia.

Retouching Reality

All of these results are kinematic, that is, independent of whatever forces may be exerted. They can be thought of as constraints on the dynamics, and therefore constraints on any physical cosmology.

Because the speed of light is the maximal signal velocity, it produces an event "horizon." No distant event can be known to us if it happened too recently for light to have reached us. This does not mean that such events cannot be incorporated into a description of space and time ("space-time"), but it does mean that such events cannot be causally related; we cannot influence them, nor they us. In chapter 14 we shall explore this idea in some detail.

Observational astronomy involves the detection of extended objects with a telescope (usually) and their recording on the retina of the eye, a photographic plate, or some electronic imaging device. If the object is moving rapidly, some quite unexpected and nonintuitive results emerge from special relativity. When relative speeds are near that of light, measured sizes and rates of expansion are far from those an observer at rest with the distant object would obtain. One of the most startling developments in twentieth-century astronomy is that the treatment of these distortions is neither trivial nor unimportant; such objects, moving relative to us at near the speed of light, or expanding at near c, are in fact seen. In attempting to unravel these extraordinary phenomena, the rules of special relativity must be applied to process or "undistort" the telescopic images. Sometimes this is straightforward, as when independent evidence of large velocities is present. In other cases the very fact of the velocity may be hidden, and analysis is hazardous.[12]

Space travelers, if we unleash our imaginations, are also governed by special relativity. At a velocity close to that of light, the fastest possible space vehicle launched from the solar system could not travel the 15,000 light-years to the near edge of our galaxy and return in less than 30,000 years, probably an intolerably long time for those who wait on earth. Time dilation could make the elapsed time *for the travelers* much shorter,[13] but to cut the ship time to 30 years would require that the average speed of the vehicle be 99.9995 percent of the speed of light. Because of the velocity dependence of inertia, achieving this speed would be a thousand times more costly than would have been computed from prerelativistic dynamics. Small wonder, then, that plausible proposals for attempting interstellar communication involve listening for electromagnetic signals, rather than sending out "space probes."

Widening Horizons

The ramifications of special relativity we have examined hardly affect our everyday existence, which is still essentially Newtonian in outlook. It is only in the extremities of the vastly large and the extraordinarily small that the conditions of relativistic velocities arise. But these are where the frontiers of man's knowledge of the universe lie. Thus, if one asks what all of this has to do with cosmology, it should be clear, or will soon become so, that the relativistic kinematics just examined become important as soon as we try to build a quantitative model of the universe. Far more important, however, is the fact that the "principle of relativity" almost begs for a generalization to all observers, whether inertial or not. Such an idea leads, in ways very far from obvious at the moment, to a geometric theory of gravity and ultimately to the tools for constructing the entire edifice of modern cosmology. Such was Einstein's originality that there is hardly a hint, in what we have said so far, of what Einstein's new theory would be.

CHAPTER 13

General Theory of Relativity

Henceforth space by itself, and time by itself, are doomed to fade away into mere shadows, and only a kind of union of the two will preserve an independent reality.[1]

Hermann Minkowski (1908)

If the two centuries after Newton represent a period of consolidation in physics, a working out of the Newtonian paradigm (to use the Kuhnian term), in cosmology they could be thought of as the search for a paradigm. In the period 1907–1915, when Einstein's ideas about the generalization of relativity were taking shape, controversy raged over the meaning of the observations, and there was only Newton's theory of gravity.

The famous Curtis-Shapley[2] debate on the nature and distance of the "spiral nebulae" took place at the National Academy of Sciences in Washington, D.C., on April 26, 1920, with Einstein in the audience, and it was at least 1926 before the work of Edwin Hubble[3] seemed clearly to establish the great distance of these objects, and the fact that they were galaxies like our own. Einstein was much more interested in the theoretical side, and in particular with what a proper theory of gravity should be like. As he indicated in his "Autobiographical Notes," he was explicitly searching for an essentially simple, elegant theory of gravitation. He used as his model thermodynamics, which stood out to him as an example of what a physical theory ought to be like. Einstein, in fact, did not wait for a resolution to the observational questions, and indeed was mostly indifferent to the details of the controversies.[4]

In the eighteenth century Thomas Wright and Immanuel Kant, among others, had indulged in far-reaching Newtonian cosmological speculation. That Kant is renowned for his systematic philosophy rather than his cosmology, and Wright is a minor figure in the history of science, that both developed largely qualitative cosmologies exposes the premature character of these attempts. The great figures of eighteenth- and nineteenth-century astronomy are observational astronomers like Flamsteed and Herschel and mathematicians like Laplace and Bessel, who developed the language of celestial mechanics. In that period, understanding of the mechanics of the solar system became almost complete; the same principles applied to binary star systems led to the knowledge of the masses of stars, and their composition yielded to the techniques of spectroscopy. Yet at the end of the nineteenth century, man looked out on a universe that might or might not be infinite, but he could claim only the most limited knowledge of it—of his own neighborhood, really, a few hundred light-years across. What we now take to be the one or two great facts about the universe were not even guessed at. There was only a vague sense of its scale. Then, by the turn of the century, observational astronomy began to marshal the tools that would place the "whole" of the universe within reach. Within two decades or so these techniques would finally show that the spiral nebulae are vast "island universes" like our own Milky Way and that they are at enormous distances from us—millions of light-years. And not long after, the universal recession of the galaxies—the "expansion of the universe"—was discovered. Almost coincidentally, it would seem, the missing paradigm, the integrating idea, was proposed: Einstein's general theory of relativity. It arose both from his special theory of relativity, whose origins we have traced without mentioning astronomy at all, and from his insights into the nature of gravity. The state of nebular astronomy, a field destined to be revolutionized by the general theory, had virtually no influence on Einstein.

We have seen that the state of electrodynamics at the end of the nineteenth century demanded a radical approach to the problem of fields in moving coordinate frames and that the experiments related to this brought an unmistakable message that something was wrong. Thus there is almost a feeling of inevitability about special relativity that emerges from a reading of the documents of the decade preceding Einstein.[5] The same cannot be said of the general theory of relativity. Many of its important aspects are without antecedents, and it is largely the creation of one mind. At the time of its genesis there was almost no empirical basis; only the problem of the advance of the perihelion of the orbit of Mercury can be said to have guided Einstein. Speculative and a priori in character, the general theory derives largely from Einstein's deeply rooted faith in a harmony in nature, and its discovery contains the elements typical of his path toward uncovering the truths of the physical world: an extraordinarily bold idea, capable of the most intricate elaboration, but in essence simple, intuitive, and above all, demanded by his feeling for the symmetry and harmony of the universe.

The general theory of relativity is essentially a theory of gravity; it revolutionized cosmology—indeed it *created* modern quantitative cosmology. Of the theory Max Born, himself one of the giants of twentieth-century physics, said, "the theory appeared to me then, and it still does, the greatest feat of human thinking about nature, the most amazing combination of philosophical penetration, physical intuition and mathematical skill."[6] Yet its impact has been very different from those of the other great revolutionary developments of twentieth-century physics, special relativity and the quantum theory. These have had wide application, while general relativity in the half-century since its inception has really been applied to only three main problems, all astronomical: the solar gravitational field, including its effect on light and the anomalous orbit of Mercury; the properties of collapsed stars; and cosmology, the motion of the galaxies. Nonetheless, general relativity is an integral part of the truly profound conceptual revolution that had its origin in the celebrated 1905 paper.

Einstein presented the theory in a series of papers beginning in 1907 and culminating in a full exposition in 1915.[7] As a theory of gravity, it makes possible for the first time a detailed mathematical description of the origin and structure of the universe. This is because the properties of the universe on the large scale are dominated by gravitational forces. There are other theories of gravitation, mostly variants of general relativity, but no competing theory has a comparable history of more than fifty years of having successfully survived repeated careful tests of its validity, nor has any other theory had anything like its impact on the development of modern cosmology.

Gravity is in an important way unique among the forces of nature, because, as Galileo showed, it acts with complete indifference on all bodies, giving them equal acceleration. Whereas Galileo's accuracy could have been no better than plus or minus a few percent, in 1890 Eötvös verified the equality of the acceleration of objects of different mass and composition to an accuracy of one millionth of one percent. Further experiments have established the equality to one part in one trillion. It is, then, an experimental fact that the force of gravity is proportional to the mass of an object and the acceleration is the same for all objects.

There are other forces known to have the same property, in particular the *coriolis* and *centrifugal* forces. But these forces (called "fictitious forces") appear when one tries to treat an accelerating (e.g., rotating) coordinate system as inertial; they are the price one pays for choosing the wrong geometry.[8] Is gravity also a "fictitious" force, an apparent force experienced only in accelerated or noninertial coordinate systems? Gravity can indeed be quite easily simulated by means of a simple acceleration. If we employ the elevator thought-experiment made famous by Einstein (fig. 13.1), we see that, to an observer inside a closed elevator, there is no obvious way to distinguish between an elevator at rest, or moving uniformly in a gravitational field (which would cause an acceleration), or an elevator accelerating upward at g in a region where there is no gravitational field. Gravity can thus be replaced by an acceleration.

An observer in a freely falling elevator or in a coasting spacecraft in a gravitational field is an inertial observer.[9] In the reference frame of the falling elevator, for example, the path of a free particle is a straight line and a ball released in such an elevator will not accelerate with respect to it but will remain suspended. Let us suppose that such an elevator is given an upward acceleration. A ball released in the elevator will now accelerate toward the floor. All objects, of whatever mass, will have the same downward acceleration. Such a behavior is characteristic of a gravitational field. Generalizing this idea, which is known as the "principle of equivalence," we could say, with Einstein, that inertial and gravitational forces are equivalent, that there is no way to tell, from the inside of an elevator for example, whether it is accelerating upward or is fixed in a gravitational field.[10] Einstein's insight was that gravity is not to be considered a force at all but in some sense a choice of coordinates.[11]

The principle of equivalence is of truly profound significance because most of the classical tests of general relativity are tests only of that hypothesis. Furthermore, it permits a qualitative description and even derivation of many of the important and dramatic consequences of the theory without a need for the elaborate mathematical framework (fig. 13.2) in which it is necessarily cast.

In Search of a Theory

As early as 1907 Einstein had become interested in the effect of gravitation on light, a problem that had attracted the attention of Newton, Laplace, and others. The principle of equivalence may be traced to that year as well. In the same year, in a review of special relativity,[12] Einstein proposed the extension of the principle of relativity to noninertial observers. The general principle of relativity or "general principle of covariance" is an extension of the postulate of symmetry from special relativity to include all observers, whatever their motions. That is to

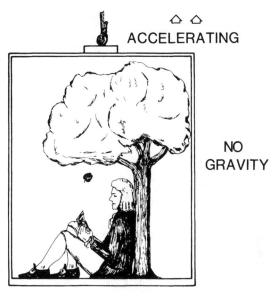

ACCELERATING

NO
GRAVITY

Figure 13.1A Einstein's elevator. The apple in the elevator is accelerating toward the scholar because it is not attached to the upward-accelerating system.

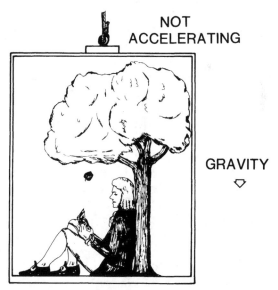

NOT
ACCELERATING

GRAVITY

Figure 13.1B The elevator is not accelerating, but is fixed (or moving uniformly) in a gravitational field so that the apple accelerates toward the scholar in the same way. The scholar cannot determine the cause of the acceleration if he restricts his measurements to the interior of the elevator.

$$R_{ik} - \frac{1}{2}Rg_{ik} = \frac{8\pi}{c^4}GT_{ik}$$

Figure 13.2   Einstein's field equations of general relativity. R_{ik} is the Einstein-Ricci tensor, a measure of the departure of the geometry of space-time from Euclidean geometry; T_{ik} is the "energy-momentum" tensor; g_{ik} is the metric tensor; R is the Riemannian curvature.

say, the laws of physics, properly formulated, are claimed to be of the same mathematical form for all observers, whether inertial or not. Einstein expressed it in this way in 1916: "The general laws of nature are to be expressed by equations which hold good for all systems of coordinates, that is are covariant with respect to any substitution whatever."[13]

Although it was to be eight years before Einstein published general relativity in its final form, and the road leading to the general theory was to be long and roundabout, he had already stated the two most important postulates of the theory in 1907. Then, in 1911,[14] Einstein first made the prediction of the magnitude of the bending of starlight by the sun, based on the principle of equivalence, but using the Newtonian theory of gravity. He was a bit apologetic about the prediction, lacking at the time a fully relativistic theory of gravitation. The predicted deflection was only one-half the value later to be calculated from full general relativity, the other half representing the contribution from the curvature of space.

During the period 1912–14 Einstein learned non-Euclidean differential geometry from the mathematician Marcel Grossmann, a colleague in Zurich and the same "friend who attended the lectures" (and who had helped obtain the position for him at the patent office in Bern) from his student days. The properties of differential geometry are expressed in the equations of tensor calculus. Einstein began work on a tensor theory of gravity, and in 1914 he and Grossmann published a paper[15] in which the basic mathematical formalism of general relativity was first presented. The theory was still far from satisfactory, and in the next two years Einstein pursued various blind alleys, at one time rejecting as nonphysical the correct field equations of general relativity. Finally, on November 25, 1915, Einstein presented the general theory of relativity in essentially its present form to the Berlin Academy.[16] Among other things, it gave a prediction for the bending of starlight near the sun of twice the Newtonian value.

In his "Autobiographical Notes" Einstein has discussed the considerations that guided him in his search. The principle of equivalence with its consideration of accelerated clocks had helped him to see that the linear mathematics of special relativity would be inadequate, and the general principle of covariance provided a filter against theories of insufficient symmetry. There are, unfortunately, still infinitely many "laws" that could be cast in covariant form. Einstein's approach was to choose the "simplest possible" systems of equations within the constraint he had set. But the very simplest equations are trivial and the correct equations must after all *describe* gravity even in the Newtonian (low-mass) limit. Of the arduous and frustrating search Einstein remarked:

In the light of knowledge attained, the happy achievement seems almost a matter of course, and any intelligent student can grasp it without too much trouble. But the years of anxious searching in the dark, with their intense longing, their alternations of confidence and exhaustion and the final emergence into the light— only those who have experienced it can understand that![17]

Mach's Principle and Inertia

Eötvös's experiment showed that to very great accuracy the inertial and gravitational masses of an object are equal. That this should be so can hardly be regarded as accidental; what is it about the inertia of a body that is related to its gravitational mass? If in the vicinity of the earth one drops a heavy mass, it accelerates downward owing to that attraction of the earth, and if it is pushed along a flat surface it requires a large force to accelerate (inertia). Removed to a great distance from any source of gravitational field, it retains its inertia but has little or no gravitational interaction. The masses manifested in these two ways, by gravitational effects and through inertia, would seem to be distinct, but experimentally they are not! Why? One answer comes from invoking what Einstein called "Mach's principle," a global hypothesis about the nature and origin of inertia.[18] Mach asserted that the source of the inertia of a body was the gravitational attraction of the distant masses of the universe. Consider, for example, the Foucault pendulum.[19] That the plane of oscillation of the pendulum is at rest with respect to the stars is a remarkable coincidence, unless an inertial coordinate system is one in which the distant stars are fixed, that is, not rotating.

Newton's famous experiment with a rotating pail of water, described in the *Principia*, provides another illustration. If we wind up a rope supporting a pail of water, then allow it to unwind, the pail begins rotating and then finally the water starts to rotate, though more slowly than the pail. The surface of the water becomes concave as a result of the rotation. Eventually the water and pail come to rotate at the same rate, so there is no relative motion. With respect to what is the water then rotating? Newton said with respect to absolute space (an answer regarded by Leibniz and Huygens as unsatisfactory). Since we now consider absolute space to be an untenable idea, what then is to be our answer? Mach argued that the water was rotating with respect to the distant masses in the universe. The properties of the solar system show that local masses— the sun, Jupiter—contribute negligibly to the inertia of the planets. Newtonian mechanics works for the solar system: a single inertial frame serves and masses do not change as the distances among the planets change. Mach's principle attributes inertia to the influence of distant masses in the universe and the inertial coordinate system of the solar system is fixed by the distant stars. From this point of view, the flattened shape of our galaxy implies that even on that scale rotation is with respect to the truly distant matter in the universe. Today Mach's principle is controversial, variously regarded as a fundamental condition of any correct theory of gravitation, as a tautology, or even irrelevant.[20]

The Structure of the General Theory of Relativity

The mathematical formalism in which general relativity is expressed, differential geometry, makes it impossible to give anything more than a general outline here. Essentially, the field equations of general relativity yield the geometry of space-time at any point if one gives as input information about the masses (and energy) present everywhere. Knowing the geometry of space-time, it is possible to calculate the path of a particle along a geodesic in curved space-time, to predict the path of a light ray, etc. Ultimately, and Einstein did this in 1917, one can express the equations in a form that is applicable to the universe as a whole, whence general relativity, if correct, becomes the framework in terms of which cosmological questions are to be posed and answered. The language of general relativity is, then, non-Euclidean differential geometry, and its essential elements are the general principle of relativity (or covariance), the principle of equivalence, Mach's principle (perhaps), and a correspondence principle that specifies the theory must reduce to Newton's in some limit. The test of these ideas is in the agreement of predicted phenomena with observation, and so we shall shortly turn to the question of experimental or observational tests of general relativity.

For the moment, however, a brief survey of some simple aspects of non-Euclidean geometry is in order.[21]

In motivating a discussion of non-Euclidean geometry, we might be pardoned for using the observation of the bending of starlight passing near the sun as justification, even though historically it emerged as a prediction of general relativity and its geometric foundations. Seen this way, the deflection of light in a gravitational field raises questions about the geometry of space.

That is, if the natural answer to the question "what is the shortest path between two points in space?" is "the path of a light ray,"[22] what happens to the concept of minimal distance between two points if the path light takes depends on the distribution of matter? It would seem impossible to find a better definition of "straight line" or *geodesic* than the path of a light ray. But if geodesics are not ordinary straight lines in the ordinary sense, then plane or Euclidean geometry cannot describe space (or space-time).

The development of geometrical systems that, while logically consistent, do not conform to all the axioms of Euclidean geometry was one of the central problems of nineteenth-century mathematics. Before 1826 Carl Friedrich Gauss, the "prince of mathematicians," developed the first propositions concerning spaces in which there are infinitely many lines through a given point, parallel to another line. After the publication of independent work of a similar nature by Nikolai Lobachevski and Janos Bolyai, Bernard Riemann in 1854 showed the existence of spaces in which *no* parallel lines can be drawn, and put into mature form the differential geometry, the grammar of general relativity.

The demotion of Euclidean geometry from its privileged status was not accomplished without the controversy that surrounds a startling, innovative idea, but by the end of the century mathematics had begun to settle into its emphatically abstract mold.

Some non-Euclidean geometries seem abstract in the extreme—geometries in which closed curves cannot be contracted to a point, multiply connected geometries (fig. 13.3); they seem obviously to have nothing to do with real space. But Gauss, in a letter to a friend in 1824, wrote, "It seems to me that we know . . . too little, or too nearly nothing at all, about the true nature of space, to consider as absolutely impossible that which appears to us unnatural."[23] And Lobachevski proposed in 1840 that astronomical tests ought to be attempted, since "There is no other means than astronomical observation for judging the exactness which attaches to ordinary geometry."[24]

Leaving aside the questions of formal validity that so vexed the mathematicians, we can learn something by examining curved surfaces. The

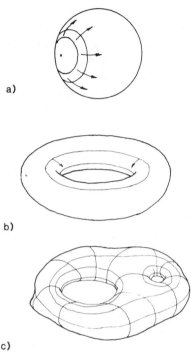

a)

b)

c)

Figure 13.3 (a) A "circle" of ever-increasing size on the surface of a sphere must eventually contract to a point. (b) Some curves can never be contracted to a point. (c) Another possible analog of the geometry of three-dimensional space.

surface of a sphere, considered as a two-dimensional space, obeys non-Euclidean geometry; for geodesics (or "shortest lines") we use arcs of great circles (fig. 13.4). On the sphere no parallel geodesics can be drawn, since every pair of great circles intersects (twice). A consequence of the postulates of Euclidean geometry is that the sum of the interior angles of a triangle is 180°; on the surface of a sphere the sum is greater than 180° and the Pythagorean theorem does not hold. The sphere is an example of Riemann's "elliptical" geometry (no parallel geodesics); the Euclidean limit is a sphere of infinite radius. The "hyperbolic" geometry studied by Gauss, Lobachevski, and Bolyai—a geometry with infinitely many parallels (nonintersecting lines) to a given line is illustrated by a two-dimensional saddle-shaped surface (fig. 13.4b). In these geometries the sum of angles in a triangle is always less than 180°.

The sphere has constant positive curvature. Hyperbolic geometries, such as the saddle have negative curvature.[25] A surface with zero curvature is "flat," that is, Euclidean. That a cylinder is a flat surface can

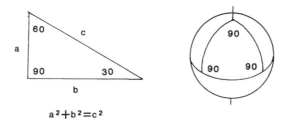

$$a^2 + b^2 = c^2$$

Figure 13.4A A spherical surface is a two-dimensional non-Euclidean "space." The rules of Euclidean (plane) geometry are replaced for the sphere by appropriate generalizations. Geodesics ("straight lines") are great circles. There are no parallel lines. The rule that the sum of the interior angles is 180° does not hold for triangles formed from three great circles on the surface of a sphere.

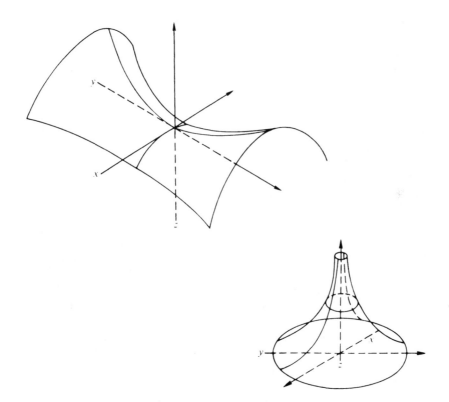

Figure 13.4B Surfaces with negative curvature. Geodesics are open. Through any point infinitely many lines "parallel to" a given line can be drawn. From Rudolf V. Rucker, *Geometry, Relativity, and the Fourth Dimension* (New York: Dover, 1977).

be seen by thinking of it as a plane wrapped on itself so that triangles and parallels are not "disturbed."[26] A surface such as a toroid (dough-nut) is more complicated, having positive curvature in some regions and negative in others.

The importance of curvature for physical applications in three and four dimensions is that measurements within the space are sufficient to determine the curvature. Much as Lobachevski suggested, a variety of measurements on a large enough scale could reveal whether space has positive or negative curvature. Adding the angles of a huge triangle enclosing the sun, for example, would give more than 180°, because light bends *toward* the sun. Space near a star or planet has, therefore, positive curvature. As we shall see in the next chapter, the large-scale geometry is determined by the distribution of all the matter in the universe. Three-dimensional spaces with negative curvature are infinite in volume, as in Euclidean (Newtonian) space, but spaces with positive curvature have finite volume. It is therefore possible, as Einstein be-lieved, that the universe is closed, finite, but unbounded.[27]

Taylor and Wheeler[28] give a graphic example of two travelers con-fined to the surface of a sphere who start off on "parallel" paths, keep-ing the same heading, (say, north) only to find after a while that they are getting closer to each other. They are, moreover, approaching each other at an ever-increasing rate. They could attribute this acceleration to some force of mutual attraction. If they found out that their space is non-Euclidean, they could as well say, as we would from outside, that the acceleration has a purely geometric origin. In the same way Einstein attributes gravity to the geometry of space-time. If the geodesic is curved compared to a Euclidean straight line at each point, the corresponding acceleration is that due to a "gravitational force."

Our discussion of curved space has understated the elegance of gen-eral relativity, because our examples give little insight into the role of *time* as a coordinate in Einstein's equations. Suffice it to say that time is "curved" from its Newtonian value. This is evident in the gravitational red shift, wherein the frequency of light is altered by a massive body. A more subtle example is that of the planets, considered as free bodies moving on geodesics in space-time. Without the sun the planets would move in straight lines (zero-mass geodesics).

Essentially this idea had occurred to Clifford, who in 1870, nine years before his death from tuberculosis at the age of 34, wrote:

I hold in fact (1) That small portions of space are in fact of nature analogous to little hills on a surface which is on the average flat; namely that the ordinary laws of geometry are not valid in them. (2) That this property of being curved or distorted is continually being passed on from one portion of space to another after the manner of a wave. (3) That this variation of the curvature of space is what really happens in that phenomenon which we call the motion of matter, whether ponderable or ethereal.[29]

What is missing from this breathtakingly far-seeing passage is the idea of *space-time*, whose importance was discovered by Einstein and Rudolph Minkowski. Gravitation is a manifestation of the deviation of the properties of real space-time from those of its Euclidean counterpart; the path of a baseball or a planet is a geodesic in space-time.

Observational Tests

The classic tests of the theory are (1) the bending of starlight, (2) advance of the perihelion of Mercury, (3) the gravitational red shift, and (4) time-delay experiments. There are other possible tests of elements of the theory, including tests of the principle of equivalence itself.[30]

Of the predictions of the general theory, the most exciting was the bending of light in a gravitational field. Following Einstein's 1911 suggestion that this could be empirically tested by observing starlight passing near the limb of the sun during a total solar eclipse (fig. 13.5), an attempt was to be made in the Crimea in August 1914 by E. Finlay-Freundlich of Berlin Observatory. Unfortunately the onset of World War I resulted in the arrest of the members of the expedition by the Russians and the impoundment of their equipment. Not until five years later was another attempt made, during the eclipse of May 29, 1919, visible in South America and Africa. Two British expeditions, one led by Arthur Eddington, obtained results entirely consistent with the theory.[31] These were announced to the world at a dramatic meeting of the Royal Society and the Royal Astronomical Society on November 6, 1919. Einstein had predicted a deflection of 1.75″, while the average observed deviation was 1.8″; confirming the predictions within the limits of experimental uncertainty, which were about 30 percent. This first striking confirmation of the general theory demolished much of the skepticism with which it had been greeted and pushed Einstein into the public limelight for the first time.

It is now possible to carry out essentially the same experiment without waiting for solar eclipses by using the techniques of radio astronomy. (Radio waves, as well as x-rays and γ-rays, are of the same nature as light, differing only in the way that sounds of different pitch do, that

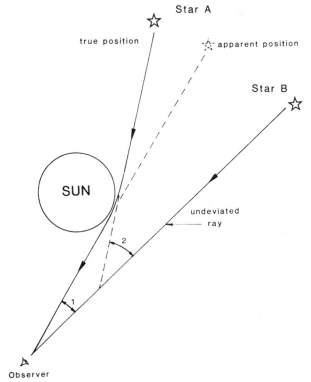

Figure 13.5 The deflection of starlight by the sun's gravitational field. The angle between star A and star B is measured during a total solar eclipse (angle 1) and is found to be smaller than the angle between the same stars (angle 2) when the sun is not nearby. The deflection is greatly exaggerated in the figure. For starlight which would graze the sun's surface the deflection would be less than one thirtieth of a degree (1.7″).

is, frequency.) The two quasars 3C273 and 3C279 are about 8° apart in the sky. Each October 8 the sun passes in front of 3C279, providing an ideal opportunity to measure the bending of the radio signal from 3C279 by comparing the apparent angular separation of the two sources as the one signal grazes the sun. Ultimately this technique will provide a very sensitive test of this prediction of general relativity. Recently additional confirmation has been obtained in a dramatic way: the discovery of "double quasars," interpreted as two images of the same quasar caused by a "gravitational lens" effect due to the bending of the radio waves from the distant quasar by a nearer object in the line of sight.[32]

A related test has become possible with the development of radar capable of reflecting signals from the planets and of interplanetary probes that can be placed into orbit about the sun or the nearby planets. This kind of experiment, first suggested by Shapiro in 1964,[33] measures the time delay experienced by a radio signal passing by the sun after being reflected from Mercury or Venus when near the sun, or by a signal transmitted by a vehicle orbiting the sun, or in orbit about a planet. The time delay is on the order of $100\mu s$ and thus easily measured. Both of these kinds of experiments have been carried out, the latter in connection with the Mariner and Viking series of Martian missions. The results are again consistent with the general relativity, with uncertainties of only 0.2 percent.

The earliest verification of general relativity involved the advance of the perihelion of the orbit of Mercury. It has been known since 1859 that the perihelion of Mercury, the point in its orbit when it is nearest the sun, advances slowly in the same direction the planet revolves so that the entire orbit rotates in space once in 23,000 years. The orbit, then, is not a closed curve, but a precessing ellipse, as shown in fig. 13.6. After the most elaborate attempts were made by Leverrier and

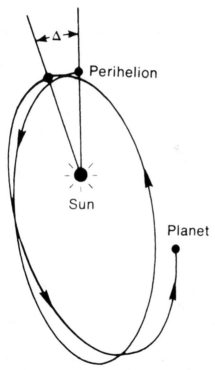

Figure 13.6 The orbit of Mercury is not quite a closed ellipse. The point of perihelion (nearest the sun) advances at the rate of about 5600" per century. After all corrections for the effects of the other planets are made, there remains a discrepancy of 43" per century which general relativity explains. W. Rindler, *Essential Relativity* (New York: Springer-Verlag, 1977).

others to explain the perihelion advance by taking into account the perturbations of the other planets and using Newtonian celestial mechanics, a discrepancy of 43 seconds of arc per century remained. Various attempts were made during the nineteenth century to remove the discrepancy. One was to postulate the existence of a hypothetical planet, Vulcan, which was supposed to orbit the sun inside the path of Mercury; another involved modifying Newton's law of gravitation. By 1907 Einstein was hoping to find a theory of gravity that would naturally explain the excess perihelion advance, and in 1915 he announced his solution, and a calculated advance of 43.03″. What little observational stimulus there may have been in Einstein's search for a theory of gravity came from this problem of the rotation of Mercury's perihelion. The effect is observable in the orbits of Venus and Earth as well, although smaller by a factor of 5 to 8, and has been measured for the asteroid Icarus. For PSR 1913 + 16, a close binary pulsar (see chapter 15), the "periastron" advance is 4° per year, or 35,000 times the effect on Mercury's orbit!

A qualitative understanding of the general relativistic correction can be obtained through the principle of equivalence. If we apply special relativity to the orbit of a planet, we find that our measuring stick will be contracted along the direction of free-fall of an inertial observer, that is, along the radius from the sun to the planet. Thus the radius of an orbit is overestimated owing to the shortened measuring rod, but the circumference, involving a measurement at right angles to the direction of free-fall, is unchanged. The result is that the circumference is less than the Euclidean value of $2\pi R$.

Applying this argument to the orbit of Mercury, we find that the distance traveled by Mercury in one period is greater than the Euclidean circumference; hence the planet executes slightly more than one revolution during one period. As a result, the point of perihelion advances from revolution to revolution by just the amount to explain the discrepancy.

Another important classical test of general relativity involves a measurement of the loss of energy of light as it "climbs" out of a gravitational well, thereby being shifted toward the red, an effect predicted by Einstein in 1911. This was first observed in the spectrum of the white dwarf companion of Sirius in 1929. The first laboratory verification of this "gravitational red shift" was provided by Pound and Rebka[34] in 1960, using a photon beam that traveled between the basement and third floor of the Jefferson Physics Laboratory at Harvard University, a distance of 22.6 meters. To understand the experiment with the aid of the equivalence principle, consider an observer in an accelerated coordinate system ("elevator") who sends a beam of photons "upward," that is, in the direction of the acceleration. The photons are detected in the accelerating system a distance away.

During the transit of the photons, the detector has gained a velocity above that possessed by the source when the light was emitted. From the *Doppler effect* the expected red shift in the Harvard experiment was about 2.5×10^{-15}, or 2 parts in 1,000 trillion! The red shift that Pound and Rebka found was in good agreement with this result. This was *the first laboratory test of general relativity.* In the case of light leaving the surface of a neutron star the shift would be about 1 percent, and when the radius of a star has a critical value called the "Schwarzschild radius" $(2GM/c^2)$, the photons are red shifted to zero frequency. In the eighteenth century Laplace[35] obtained this very result by considering the mass required for the escape velocity from the surface of a star to equal the velocity of light. That no radiant energy can escape from a sufficiently strong gravitational field is in essence a prediction of the existence of "black holes," whose detailed properties will be examined in chapter 15.

Conclusion

General relativity ascribes gravitation entirely to the geometry of space-time, determined on the small and large scale by the distribution of matter in the universe. The path of a free particle is a "straight line," a geodesic, but in a gravitational field, the geodesic is "curved"; that is, the geometry is non-Euclidean. The principle of equivalence, the statement that a gravitational field is indistinguishable from the effects of an accelerated coordinate system, leads to this interpretation of gravity as an acceleration deriving from the curvature of space-time. Yet the theory, more than 60 years after its birth, is still not fully tested. Its predictions are under constant scrutiny, and its success is continually being reevaluated as measurements become more precise or as new experiments become possible. There are competing theories, such as the scalar-tensor theories of Jordan, and later Brans and Dicke. Moreover one expects no greater degree of permanence for general relativity than for other theories of comparable success and scope. It may very well be that it will come to be seen as a first step toward a more nearly complete and comprehensive theory.

For example, at extremely short distances, on the order of 10^{-33} cm, the geometry of space is subject to *quantum fluctuations,*[36] and even the concepts of space and space-time have only approximate validity. Thus, in the final analysis, general relativity and the quantum theory must be reconciled. It bids fair to be a protean task, worthy of the best minds of our time. From this program will surely come an understanding of space and time that resembles our present ideas only in their broadest outlines. In this regard, one can do no better than to quote John Wheeler, one of the most imaginative physicists of this generation:

No point is more central than this, that empty space is not empty. It is the seat of the most violent physics. The electromagnetic field fluctuates. Virtual pairs of positive and negative electrons in effect are continually being created and annihilated, and likewise pairs of mu mesons, pairs of baryons and pairs of other particles. All of the fluctuations coexist with the quantum fluctuations in the geometry and topology of space. Are they additional to those geometro-dynamical zero-point disturbances or are they in some sense not now well understood mere manifestations of them? . . . What else is there out of which to build a particle except geometry itself? And what else is there to give dis-creteness to such an object except the quantum principle?[37]

The theory, then, is in some sense only in its infancy. As with any theory, the test lies in the comparison with observation or experiment. When high-energy physics begins to probe the domain where quantum fluctuations in the geometry of space become important, if not before, general relativity will likely undergo dramatic change, But whatever its future, Einstein's theory has forced the discarding of older ideas about space and time and provided a profoundly revolutionary view of the universe in which we live.

CHAPTER 14

Primeval Atom

The evolution of the universe can be compared to a display of fire-works that has just ended: some few red wisps, ashes, and smoke. Standing on a cooled cinder, we see the slow fading of the suns, and we try to recall the vanished brilliance of the origin of the worlds.

Georges Lemaitre, 1931

Foundations of Modern Cosmology

By 1915 Einstein's theory of gravitation had largely brought to completion the revolution in ideas about space and time that is the major legacy of the relativity theory. Yet with the theory still in its infancy and virtually untested, Einstein broke new ground by applying general relativity to the problem of the structure and evolution of the universe. Because the large-scale features of the universe are due to the action of gravitational forces, general relativity, Einstein knew, would provide a framework in terms of which one might endeavor to understand these features. So it was that in 1917 he first examined the cosmological implications of the general theory.

It is crucial to an understanding of the early attempts at a quantitative cosmology to appreciate the context in which they were made and the state of observational knowledge of the universe before 1920. In 1912 Henrietta Leavitt at Harvard College Observatory discovered the relationship between the intrinsic brightness and period of pulsation of the Cepheid variable stars that made possible, in the ensuing years, the determination of the distances to nearby galaxies (the "Cepheid variable yardstick"), but it was only in 1929 that the universal recession of the distant galaxies was established by Edwin Hubble,[1] and as much as a quarter century later this was still a matter for debate.

Our present picture of an expanding universe whose "elementary particles" are galaxies or "island universes" (fig. 14.1) like our own Milky Way galaxy and whose velocity of recession is proportional to the distance from us thus dates from only about 1930 and has become generally accepted only since the Second World War.[2] So it was reasonable in 1917 for Einstein to assume that the universe was static on the large scale, neither expanding nor contracting. To obtain a static solution to the field equations of general relativity, he had to modify them by adding a new term, the so-called "cosmological constant," which in effect introduced a repulsion (or attraction) at large distances and permitted the desired static solution. Einstein closed his 1917 paper[3] by saying of the cosmological constant, "that term is necessary only for the purpose of making possible a quasi-static distribution of matter, as required by the fact of the small velocities of the stars." The resulting solution described a closed and finite but unbounded universe whose dimension (radius of curvature) could be estimated on the basis of a rough idea of the amount of matter in the universe. To Einstein, the strongest appeal of a finite universe was its mathematical simplicity.

Shortly after Einstein's initial application of the general theory to the structure of the universe, Willem de Sitter obtained another static solution, for a universe devoid of matter. This discovery that there can exist curved space devoid of matter was at first of interest only to mathematicians, but de Sitter's work gave impetus to the search for additional solutions to Einstein's equations.[4] In the 1920s the Russian meteorologist and aviator Alexander Friedmann (1922), the Belgian Abbe Georges Lemaitre (1927), and the American Howard Robertson (1928) found general *nonstatic* solutions. These models of a homogeneous and isotropic[5] universe continuously expanding or undergoing cyclic expansion and contraction represent the culmination of one of the great achievements of theoretical physics: the prediction, before it was observed, of the expansion of the universe.

By 1929 the velocity–distance relation had been established by Hubble and others; it was found that the light of *all* distant galaxies was shifted toward the red (fig. 14.2) and by an amount that depended on distance. Because the red shift was immediately interpreted as a Doppler shift[6] due to recession, it followed that, the more distant the galaxy, the faster it was receding! (This does not mean that as a galaxy recedes from us, it speeds up; depending on the model, it may mean the opposite.) Because of its astonishing implications the Doppler interpretation of the galactic red shifts was debated at length and other explanations were proposed. The reddening effect of interstellar absorption was raised as a possible cause, as were "tired light" theories, which imagined that photons might somehow age while traveling for hundreds

Figure 14.1 The Coma cluster of galaxies. This cluster, located at a distance of about 350 million light years in the direction of the constellation Coma Berenices, contains over 1000 members. Most of the objects in the photograph are galaxies containing billions of stars each. Note the stellar image at center left. Courtesy of Kitt Peak National Observatory.

of millions to billions of years from a distant galaxy to the earth. In the end, however, there could be little doubt of the universal recession of the galaxies.[7] The most persuasive evidence for the red shifts as a velocity effect comes from two facts: that the age of the universe derived from a backward extrapolation of the expansion is consistent with that obtained by entirely different means (age of the stellar clusters, the earth's

RELATION BETWEEN RED-SHIFT AND DISTANCE
FOR EXTRAGALACTIC NEBULAE

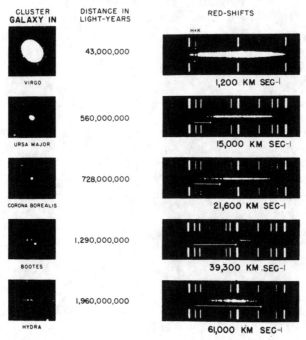

CLUSTER GALAXY IN	DISTANCE IN LIGHT-YEARS	RED-SHIFTS
VIRGO	43,000,000	1,200 KM SEC-I
URSA MAJOR	560,000,000	15,000 KM SEC-I
CORONA BOREALIS	728,000,000	21,600 KM SEC-I
BOOTES	1,290,000,000	39,300 KM SEC-I
HYDRA	1,960,000,000	61,000 KM SEC-I

Figure 14.2 Spectral red-shifts of distant galaxies (or clusters of galaxies) deduced from the Calcium H and K absorption lines. Distances based on a Hubble constant of 100 km/secMpc. Palomar Observatory Photograph.

age from radioactive decay, the time since nucleosynthesis), and the discovery of the cosmic microwave background (see below).

Does this observed recession of distant galaxies in all directions signal a return to a form of geocentrism? That it does not can be seen by considering a "raisin-bread" model of the universe (fig. 14.3). If the scale of the universe (the rising dough with raisins embedded) increases, all the galaxies recede from each other; there is no center! The same recession is seen by all observers, which is what is known as the "cosmological principle": the universe looks the same to all observers (spatial homogeneity). Nicolas of Cusa and Lucretius were right,[8] and in a sense, more so than Copernicus. Nonetheless, this idea, in slightly different form, is known as the "Copernican principle": we do not occupy a privileged position in the universe (nor does any other observer).[9]

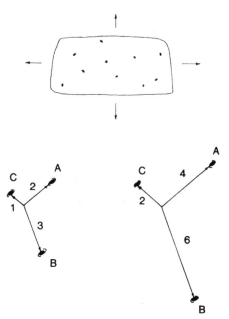

Figure 14.3 "Raisin-bread model" of the expansion of the universe. Between the two figures, the scale has increased by a factor of 2. We note that the velocity is proportional to distance: for A, v = distance/time = 4-2/1 = 2, for B, v = 6-3/1 = 3, and for C, v = 2-1/1 = 1.

Cosmological Models

The Hubble diagram of figure 14.4 shows the proportionality between the distances to a remote galaxy and the amount by which its light is shifted toward the red. With the Doppler interpretation, the same linear relation applies to the velocities of recession, galaxies ten times further away receding on the average just ten times as fast as galaxies at a given distance away: $v = Hr$. Here H is the "Hubble constant," the measure of how fast the universe is expanding at the present moment and in many cosmological models an indication of the "age" of the universe. Hubble's value was 550 kilometers per second per megaparsec,[10] nearly ten times the currently accepted value of 50–100 km/sec·mpsc. $H = 50$ means that a galaxy at a distance of 1 billion (10^9) parsecs (more than 3 billion light-years) will be receding at $50 \times 1000 = 50,000$ km/sec, or one-sixth the speed of light. An obvious conclusion

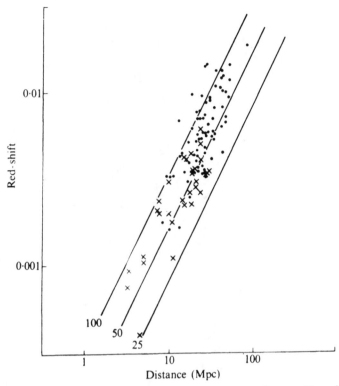

Figure 14.4 The Hubble diagram. Red-shift is plotted versus distance. The solid lines are fits corresponding to various values of the Hubble constant. Apparently a value of 50–100 km/secMpc will fit the data. C. Rowan-Robinson, *Cosmology* (London: Oxford, 1977).

from the Hubble relation is that the galaxies must have been much closer to each other in the past. The time since all the matter of the universe would have been together, gravitational deceleration being ignored, is just the reciprocal of the Hubble constant. A value of $H = 50$ km/sec·mpsc gives $1/H = 20$ billion years. A conservative value for H would be 75 ±25, and a Hubble time of 10 to 20 billion years. Because of the mutual attraction of matter, the time since the "Big Bang" (see below) is less than this value (except for some gravitational models with positive cosmological constant), by an amount that depends on the model used (fig. 14.5).

The galaxies, however, are not flying apart in the usual sense. Rather they are at rest (except for small random motions) in expanding space. The relationship between red shifts and velocity of recession depends

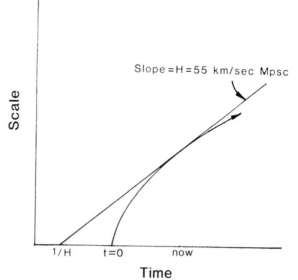

Slope =H =55 km/sec Mpsc

Scale

1/H   t=0   now

Time

Figure 14.5A The expansion of the universe displayed as a scale factor ("size of the universe") plotted against time. A popular value for the rate of expansion, $H = 55$ km/secMpc is assumed. Extrapolating backwards, this line gives the time $t = 1/H$ for the big bang, assuming a constant rate of expansion. Clearly this figure is greater than the age of the universe if the expansion is slowing down.

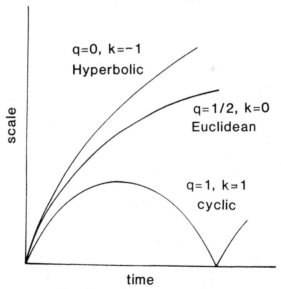

q=0, k=−1
Hyperbolic

q=1/2, k=0
Euclidean

scale

q=1, k=1
cyclic

time

Figure 14.5B The expansion of the universe for different values of the deceleration parameter q. The hyperbolic universe is open and infinite, expanding forever. The Euclidean universe is flat, but open. The cyclic universe is closed. For the latter, the cycle time is given by $T = 2\pi \, q/ \, H(2q-1)^{3/2}$, which for $q = 1$ and $H = 75$ gives 80 billion years.

on the cosmological model chosen, so that the Doppler interpretation is actually misleading.[11] The radiation is propagated through expanding space, which stretches the wavelengths.

In attempting to describe the time dependence of the scale of the universe, we must take careful note of the fact that definitions of distance and time intervals are at the heart of general relativistic cosmologies. As an illustration, we consider the "Milne" model, a simple universe consistent with special relativity, which expands at a constant rate, with zero acceleration, and has infinite space with negative curvature (fig. 14.6). Since the general relativistic curvatures in our neighborhood are small (mass density very low), this model is a reasonably good approximation to the observable universe out to distances of a few billion light years. Uniform expansion means that the cosmological distance scale, the "size of the universe," is proportional to the time elapsed since the expansion began. In any model, the relation between the red shift z [12] and the scale factor R of the universe at the times of emission and observation is $1 + z = R_{obs}/R_{em}$. In the Milne model, the ratio of the age of the universe at the time of observation ("now") to the age at emission has the same ratio, $1 + z$. Thus an enormous cosmological red shift of, say, 100 means the light left the source when the universe was 100 times more compact and only 1 percent as old as it is now. The most distant galaxies observed have red shifts of about $z = 0.5 (v = 0.4c)$, while some quasars have red shifts as high as 3.0, meaning that an absorption line in the spectrum of such an object with a wavelength of 2000A (in the ultraviolet) would appear in the infrared, having been red shifted all the way across the visible spectrum (which is 4000–7000 angstroms). Using the Doppler formula, this would represent a recession of 90 percent of the speed of light, if quasar red shifts are cosmological.[13] The present distance would be about 15 billion light-years (for $H = 75$ km/sec·mpsc), even though the emission distance was under 4 billion light-years.

Most of the acceptable models show an expansion that slows with time, although figure 14.6, which exhibits most of the plausible cosmological solutions (omitting primarily those that are uniformly collapsing and clearly not capable of describing our universe), shows that some, with positive cosmological constant, accelerate instead. The cosmological solutions to the equations of general relativity can be classified according to the curvature k ($\pm$,0) (discussed in chapter 13), the cosmological constant λ,[14] and the density ρ, usually expressed in terms of the critical density (ρ_c) required to yield a closed universe with $\lambda = 0$. Thus if $\Omega = \rho/\rho_c = 1$, the density is just sufficient to close the universe, while $\Omega < 1$ describes an open universe. Of course, initial conditions must be specified as well. In the "standard model," the universe is as-

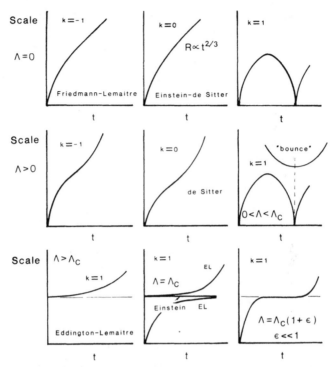

Figure 14.6 Expanding universes, classified according to the curvature k and cosmological constant Λ. Universes with $k = +1$ are finite; the others are infinite. The critical value Λ_c (the "Einstein value") is the value of the cosmological constant which yields a static universe. If Λ is at or near Λ_c, exotic expansions occur (Bottom row).

sumed to be expanding from some dense (perhaps singular) state—the primeval fireball or "big bang" (a term coined by Fred Hoyle)—but if we allow universes that originate with a "whimper," the spectrum of solutions is as given in the figures.

In one sense, the crucial question of modern cosmology is "will the universe expand indefinitely, becoming thinner and thinner (until 'the lights go out'), or will it someday recollapse?" Among the cosmological models consistent with general relativity, those that begin with a singular event or "primeval atom" seem most likely to provide a realistic description of our present universe and permit framing this question in a quantitative fashion. As we shall see, what we know of nucleosynthesis, the creation of the elements, together with the discovery of the cosmic background radiation, gives us considerable confidence in these "big bang" or singular cosmologies. Indeed, they are often described by the term "the standard model."

Table 14.1 lists the properties of various singular models with different values of these parameters λ and k. As the table indicates, the solutions can be characterized as "expanding" or "oscillating." This division depends on whether the separation R between two otherwise stationary ("comoving") points increases without limit or reaches a maximum and decreases. In figure 14.6 are represented nine different expanding universes, all but two of which begin at a zero-radius singularity. Most of these models are associated with the name of one or more of the great cosmologists of the 1920s. There are relatively fewer varieties of the oscillating universe. Since we are on the expanding segment of whatever model is appropriate, we would look back toward a big bang in an oscillating universe or in a continuously expanding one. In a cyclic universe, however, a day would come when collapse would set in. The change would be signaled by a reversal of the cosmic red shift to a blue shift, as matter gathered itself for the great infall of an inverse big bang or "big squeeze." The cosmic microwave background (see below) would begin to rise in temperature. Whether the universe could pass through the zero-volume singularity to be reborn into the next cycle is not known.

If figure 14.6 displays a surplus of universes, the picture can be made clearer by focusing on the models with zero cosmological constant favored by most cosmologists. In that case, the expansion is slowing down in all models. It is convenient to introduce a deceleration parameter as a measure of the rate of slowing of expansion. The possibilities can be classed as follows (see fig. 14.5B):

q (Deceleration) Parameter	k (Curvature) Constant	Universe
Greater than zero	−1	Hyperbolic; open and infinite (Friedmann-Lemaitre)
½	0	Flat; open and infinite (Euclidean)
Greater than ½	+1	Elliptical; closed and finite (Lemaitre)

The decision among these three possibilities can in principle be made directly from observation (fig. 14.7). Galaxy counts of various sorts promised even in the 1930s to reveal the curvature of space, but this promise has not been realized. All the source counts depend in one way or the other on the fact that the volume-to-radius ratio depends on the curvature. If the galaxies are evenly distributed throughout space, then a uniform sample would show an excess at large distances if the curvature is positive, a deficiency if the curvature is negative.[15] The problem in early counts of nearby galaxies was that the curvature distinction is significant only at large distances, and clustering obscures the effect.

Table 14.1. Properties of Various Singular Models

Cosmological Constant (λ)	Curvature (k)			"Cosmic Force"
	Negative (-1)	Zero 0	Positive ($+1$)	
Negative	Oscillatory	Oscillatory	Oscillatory	Attractive
Zero	Expanding	Expanding	Oscillatory	None
Positive	Expanding	Expanding	Oscillatory or expanding	Repulsive
Universe:	Open and infinite	Euclidean infinite	Closed and finite	

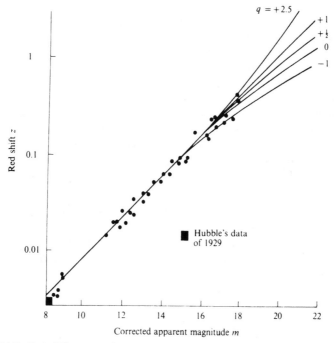

Figure 14.7 Red-shift magnitude curves for various values of the deceleration parameter q. The solid dots represent the data for the brightest members of 42 clusters of galaxies. Note that only at great distances could such a diagram distinguish between an open or closed universe. M. Berry, *Principles of Cosmology and Gravitation* (Cambridge: University Press, 1976).

As samples of enormously distant, extremely faint galaxies have become available, a new problem has arisen, namely, that the galaxies in these counts are billions of light years away and may have been substantially brighter (or perhaps fainter) when the light we see left them. The samples are consequently skewed by evolution and the (substantial) effects of deceleration and curvature are still masked.

It is nonetheless fascinating that only two numbers, the Hubble constant and the deceleration parameter, determine the geometry of the universe. In an isotropic, homogeneous universe the density is directly related to the deceleration parameter, because if there is sufficient mass in the universe, the expansion will ultimately reverse itself, and the universe will collapse. Though some of the most recent and painstaking attempts[16] to determine q using distant optical and radio galaxies have indicated a value of q close to 1, the evolutionary corrections that must be made for these objects we see as they were 5 to 10 billion years ago are largely unknown, so much so that it is probably hopeless, given our present knowledge, to decide the question on the basis of galaxy counts. The weight of current evidence from density determinations, while not yet conclusive, leans in the other direction, toward $q = 0$.[17] On the basis of currently detectable matter, the density of the universe is about 3×10^{-31} g/cm³, compared to a critical density of 4×10^{-30} g/cm³ (about two hydrogen atoms per cubic meter) required to "bind" the universe, to keep the matter from endlessly flying apart. Thus less than 10 percent of the required mass seems to be present. On the possible forms the "missing mass" might take, if indeed it is there, we shall have more to say later. Consideration of the observed abundance of deuterium ("heavy hydrogen," with a nucleus having a neutron as well as a proton) also indicates an open universe (fig. 14.8). Deuterium nuclei are quickly "eaten up" by reacting with protons to form He^3, and if one assumes that most of the deuterium we now find was made in the big bang, then the amount now remaining is a function of the density of protons (hydrogen nuclei).[18] This measurement has the great advantage that it can be made "locally"; it does not involve detection of objects at the "edge" of the universe. As Gott et al. wrote, "The objections to closed universes are formidable but not fatal; a clear verdict is unfortunately not in, but the mood of the jury is perhaps becoming perceptible."[19]

If q had the value unity, then the resulting density ($\rho = 3H^2/4\pi G$) would need to be more than ten times the known luminous density. Thus either q is much smaller, near zero, or most of the mass in the universe is invisible. Some of the mass is difficult, even impossible, to detect; recently some galaxies, including our own, have been found to

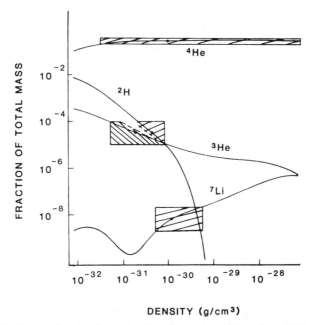

Figure 14.8 Curves showing the production of deuterium (^{2}H), helium-3 (^{3}He), helium-4(^{4}He) and lithium-7(^{7}Li) in the big bang, as a function of the present density of the universe. The observed values are indicated by the crosshatching. A value near 10^{-31} g/cm^3 seems to fit the observations. The mass fraction of ^{4}He is very sensitive to the expansion rate during the first few minutes. After C. Rowan-Robinson, *Cosmology* (London: Oxford, 1977).

have halos that may contain more matter than the previous known mass of the entire galaxy. Some evidence of intergalactic gas has been found. The fact that clusters of galaxies, which ought to be bound systems, seem not to have enough matter in them again suggests that some mass is eluding us. Undoubtedly some of the matter of the universe is hidden in the form of undetectable black holes, though how much we do not know, and there may be dead, rocky, planet-like bodies, cold and unobservable, scattered through the universe. Conceivably there are "dead" galaxies in which star formation ceased so long ago that the stars are no longer luminous, except for low-mass red dwarf stars that we would have great difficulty detecting. Some of the mass is in the form of the microwave background, but we know how much that is, and some is surely in the form of gravitational radiation (a novel prediction of the general theory of relativity), but that is not likely to be a large fraction of the mass of the universe. Finally, if neutrinos turn out to have a small but finite mass (perhaps 30,000 times less than that of the electron),[20] they could represent the dominant form of matter in the universe. The observed He[4] abundance constrains the amount of matter that could be, or could have been, in the form of protons or neutrons, so that black holes, dead galaxies, and so on, are not likely to supply the "missing mass." Massive neutrinos might or might not add enough mass to bind the universe, but certainly *could* be the answer. On the other hand, one might reasonably argue, with the present evidence indicating 10 percent or less of the required mass, that the most economical theory would be one that admitted that the density was most likely less than the critical density and that the universe is therefore probably open.

Cosmic Microwave Background Radiation

In 1965 during tests of an early satellite communications system, Arno A. Penzias and Robert Wilson discovered the earth is bathed in radio energy from the universe, coming to us with equal intensity in all directions. This *isotropic* background radiation has the spectrum of a black body[21] (fig. 14.9) with a temperature of about 3 K and is evidently of extragalactic origin. Efforts to explain the radiation as being due to some collection of discrete sources have failed, and it is generally believed that this is relict radiation, a remnant of the radiation from the primeval fireball, which has cooled after 10^{10} years to about 3 K as the universe has expanded. If we assume this to be the case, it is a most powerful argument for the big-bang or explosive theories of the formation of the universe and can be regarded, with the Hubble expansion, as one of the two great facts of observational cosmology. That such a cosmic background radiation should exist was predicted by George Gamow[22] in 1948; it is a consequence of all singular general relativity cosmologies. Before the temperature of the fireball had dropped to about 10,000 K, the universe was in a *plasma* state, and it was only after this era, about 10^{12} sec (100,000 years) after the big bang (when the red shift, z, was 1,000), that the protons rapidly combined with electrons to produce neutral hydrogen, allowing the radiation to decouple from the matter. The radiation has subsequently cooled (much like the expansion of a gas) as the universe has expanded and is now at a temperature of less than 3 K. Beyond this, we can look back no further in time![23] Of all the photons in the universe, 99.99 percent are in this microwave background. Recently, Richard Gott[24] has proposed a cosmological model in which the microwave background is "Hawking radiation" from an "event horizon" (see chapter 15). This model envisions the creation of many noncommunicating open universes as "bubbles" in de Sitter space.

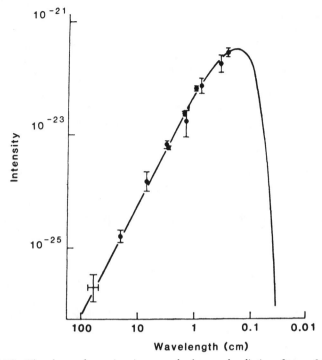

Figure 14.9 The observed cosmic microwave background radiation, fit to a 2.7K black-body curve. The units of intensity are watts cm^{-2} sr^{-1} Hz^{-1}.

Other Theories of Gravitation Cosmology

In the early years after the discovery of the universal expansion, es-timates of the size and hence of the age of the universe tended to be substantially less than the age of the earth as inferred from radioactive decay of heavy metals. This contradiction, long since resolved, pro-vided one of the motivations for the development of the *steady-state model*. As the name implies, the universe pictured in this model is one that always looks the same in time, on a sufficiently large scale. This premise, which avoids the singular creation that some cosmologists find unaesthetic, can be seen as a generalization of the "cosmological prin-ciple." As a statement of the assumption of the spatial homogeneity of the universe, the cosmological principle is plausible and immensely simplifying. The *perfect cosmological principle* generalizes this idea to in-clude time: the universe looks the same to all observers at any time whatever. The universe does not evolve in time but always has the same large-scale structure. The consequences of incorporating the perfect cosmological principle are immediate and profound. Granted the expansion of the universe,[25] the average density must decrease with time as the galaxies move away from one another, and this violates the perfect cosmological principle. The steady-state theory confronted this obvious difficulty by abandoning mass-energy conservation and postulating the continuous creation of matter at just the rate required to keep the density constant. Lest one imagine that this violation of energy conservation should provide an easy test of the theory, it turns out that, in the steady-state universe, matter must be created at the rate of about one neutron or proton per cubic centimeter per 500 trillion years. Clearly the steady-state cosmology cannot be rejected on the grounds of violation of an empirically verifiable law of energy conservation.

The idea of continuous creation is an old one, having first been suggested in 1918 by MacMillan as a way of resolving Olbers' paradox (see below). Moreover, there is another salutary effect of the invoking of continuous creation, the avoidance of the problem posed by the application of the second law of thermodynamics to the expansion of the universe. The second law of thermodynamics asserts that, in spite of energy conservation, the amount of energy available in the universe to do work never increases and in fact usually decreases. Put another way, disorder is continually increased by physical processes. The universe is "running down," tending toward a state of constant temperature.[26] In the steady-state universe the high-entropy radiant energy lost through the expansion is offset by the continuous creation of energy at low entropy.

For more than a decade the steady-state theory provided a viable alternative to the relativistic big-bang cosmologies as a basis for interpreting the observational data. Although today it has few adherents, it was an important stimulus to observation and theory. The discovery of strong evolutionary effects among galaxies and quasars and of the 3 K cosmic microwave background radiation sounded the death knell for the steady-state theory. Some cosmologists abandoned it with great regret. Dennis Sciama was one: "I must add that for me the loss of the steady-state theory has been a cause of great sadness. The steady-state theory has a sweep and beauty that for some unaccountable reason the architect of the universe appears to have overlooked. The universe in fact is a botched job, but I suppose we shall have to make the best of it."[27]

For some cosmologists the perfect cosmological principle has been replaced by the anthropic cosmological principle,[28] which states that we occupy a very *special* point in time. If it were much earlier, galaxies and stars would not have appeared or intelligent life had time to evolve. Much later, either the singularity in our future would have extinguished all life, if the universe is closed, or proton decay and the law of entropy would have robbed the open universe of all structure. Thus the universe as we see it now is *not* typical!

The steady-state theory is only the most famous of the cosmological theories that have confronted general relativity. Among the most important and most interesting is a theory attributed to Robert Dicke and Carl Brans, the Brans-Dicke or scalar-tensor theory, which is based on some earlier ideas of Pascal Jordan. The essential novelty of this approach is the introduction into the tensor field equations of a scalar field. Although a discussion of the scalar-tensor theory would fall outside the scope of this book, an important consequence of the theory is a gravitational constant G that varies with time. Physical theories in which the fundamental constants of nature, such as the velocity of light or the mass of the electron, might be time dependent had considerable currency in the 1930s and have been championed by such eminent physicists as P. A. M. Dirac, one of the founders of the quantum theory. A varying gravitational constant would manifest itself in contraction or expansion of the earth, and hence continental drift, in changes in the rate of stellar evolution, to name only two examples. As of this writing, the Brans-Dicke theory is still (barely) an acceptable alternative to Einstein's general relativity, and there is considerable interest in trying to put an upper limit on the time rate of change of the gravitational constant.

Horizons

One important consequence of the finite velocity of light is that we cannot communicate, at any given time, with the entire universe; there is a horizon beyond which we cannot see. In figure 14.10 we introduce the space-time diagram and the idea of a "light cone," which partitions space-time into the past, future, and a region that is not causally connected to us. In this diagram, light rays travel at 45° angles, and not only can one not travel backward in time, one cannot reach, or communicate with an observer outside one's forward light cone. To do so requires traveling faster than light. This means that we can "see," that is, receive signals from, only events in our past or backward light cone. Only a fraction of the universe is visible to us. The figure shows how other parts of the universe come into view (from "over the horizon") as time passes.

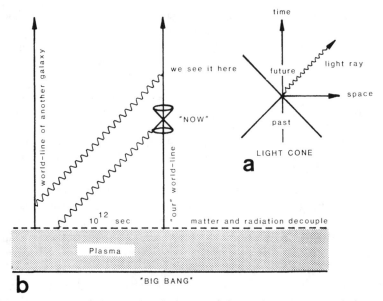

Figure 14.10 (a) A "light cone." Light rays travel along 45° paths ($x = ct$, with the speed of light c set equal to 1 for convenience). In spacetime, an object "exists" everywhere along its worldline. There is no "flow" of time.
(b) Our worldline and that of a distant galaxy not yet visible to us on a spacetime diagram. Only at some time in the future will we begin to receive photons from the other galaxy.

Note that, because stars have finite lifetimes, on the order of 10 billion years, we receive photons, at any moment in time, from only a time-slice some 10 billion years in thickness (figs. 14.11 and 14.12). This fact provides a resolution of a famous paradox raised by Cheseaux in the eighteenth century (and by Halley even earlier) and by Olbers in the nineteenth, known as Olbers' paradox, which can be expressed in the form of the question, "why is the night sky dark?" Lest one think that the answer is obvious, we should explain that no plausible answer was advanced before the advent of general relativity (the idea that the sky is dark at night because of the expansion of the universe), and the final answer given above dates from the 1960s.[29]

Quantum Cosmology

We have already seen that the observed abundance of deuterium is a function of the density of the universe and thus in principle provides a

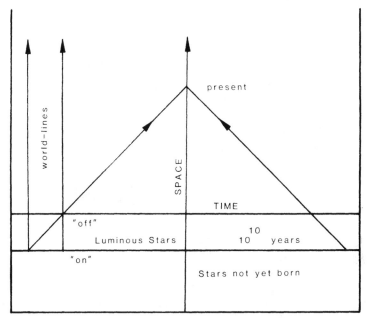

Figure 14.11 A spacetime diagram showing light rays reaching an observer from distant points in spacetime. Only those stars that were luminous during the time-slice of about 10 billion years "thickness" (a typical stellar lifetime), will be seen at any given time ("now") by the observer. After E. R. Harrison, *Physics Today*, February 1974.

way of determining whether the universe is open or closed. The primordial He^4 abundance is also a measure of the rate of expansion of the universe. The present and future properties of the universe are a direct consequence of the kinds and numbers of particles that dominated the first few minutes of its history (figure 14.13).

This brings us to the remarkable story of the connection between the microscopic world of elementary particles, and the universe on the largest scale. Beginning in the mid-1960s particle physicists began to win the struggle to which Einstein had devoted his later scientific life, that of the unification of the forces of nature. The result is a picture of three of the four forces, electromagnetic, weak (these two combined as the "electro-weak" interaction), and strong,[30] unified by a single theory of all nongravitational interactions. Of course Einstein's goal was to unify gravitation and the other forces and to give a geometrical interpretation to these interactions as he had done with gravitation. But what seems to be emerging is that all the *nongravitational* forces can be expressed in terms of a single fundamental number, the so-called "fine structure constant" and a single global theory. If so it is one of the great achieve-

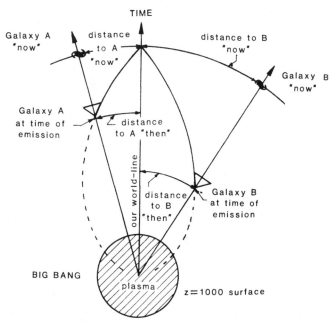

Figure 14.12 Spacetime diagram showing three observers (galaxies) in an expanding universe. Before $z = 1000$ ($t = 10^{12}$ secs) the primaeval fireball is opaque. Note how times and distances of emission and reception of a signal are related. After J. Silk, *The Big Bang*, and E. R. Harrison, *Cosmology*.

ments of physics, certainly the greatest achievement of the quantum theory. Each of these interactions is mediated by a characteristic particle, the photon in the case of the electromagnetic interaction, an "intermediate vector boson" in the case of the weak interaction, and a particle called the "gluon" in the case of the strong interaction. These grand unification theories, which are the result of the efforts of many physicists but especially of the Nobel laureates Murray Gell-Mann, Steven Weinberg, Abdus Salam, and Sheldon Glashow, are called "gauge theories." In the case of the strong interaction, they are built on the discovery that what had been thought to be the elementary constituents of the nucleus, the proton and neutron, are themselves composed of fundamental objects called "quarks" (from James Joyce's *Finnegan's Wake*). The quarks apparently occur in six "flavors," labeled u, d, s, c, b, and t, for (somewhat whimsically) "up, down, strange (or sideways!), charmed, bottom (or beauty), and top (or truth)." They are bound together in triplets to form particles like the neutron and proton and in pairs (quark-antiquark) to form certain other strongly interacting par-

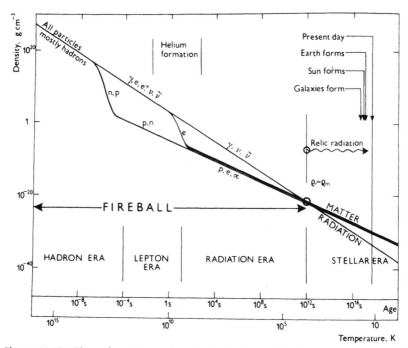

Figure 14.13 The early universe, showing that radiation dominates over matter until about 10^{12} seconds (100,000 years) after the big bang. J. Kleczek, *The Universe* (Amsterdam: Reidel, 1976).

ticles called mesons. The binding is carried out through the *exchange* of the gluons, which come in three "colors."[31] These names are not to be taken literally, of course; they simply *label* the properties ("degrees of freedom") of these fundamental constituents of matter.

Much is yet to be learned about these unified interactions and the particles responsible for them (conceivably the so-called "fundamental" particles of the weak interaction—the electron, muon, the neutrinos, and perhaps quarks and gluons as well—will turn out to be composite themselves), but the present situation represents an enormous triumph. In any case, if the theory is correct, certain very definite consequences of great significance to cosmology immediately follow.[32]

It is essential to understand how important a role symmetry and invariance principles play in this grand unification, especially since we have already made much of the role of ideas of symmetry and harmony in the thinking of such pivotal figures as Pythagoras, Kepler, and Einstein. Here the symmetry is expressed as "gauge invariance," which (coupled with what is called "spontaneous symmetry breaking") leads to a number of important experimental predictions, not the least of which concerns the existence of certain fundamental particles. All the nongravitational forces are assumed to be due to the "virtual exchange" of "vector bosons," the photon, the $W\pm$, Z°, and the gluons. The octet of colored gluons bind the "quarks" together. The details of the grand unification scheme are far beyond the scope of this work, but an important feature is the use of "Lie group theory," a branch of mathematics that deals with classes of symmetries. When applied to physical situations these ideas lead not only to conservation laws but to decay rates, mass formulas for elementary particles, and so on. Symmetry principles are thus at the heart of modern particle theory.

What concerns us here, of course, is how all of this bears on cosmology. The answer is to be found in a study of the early universe. Our current understanding of elementary particle physics when combined with the most appealing gauge theories suggests that there are six distinct light particles, called leptons ($e\pm$, ν_e, $\mu\pm$, ν_μ, $\tau\pm$, ν_τ), and six kinds (or flavors) of quarks. The present density of the universe may in fact fix this number at no more than six. Cosmological considerations then place upper limits on the masses of the two or three types of neutrinos (which may, of course, have no mass at all) and indeed place better limits than those that can now be obtained in the laboratory. From another point of view, if the masses are above some minimum, the universe *must* be closed; the missing mass has been found! These theories predict that one of the hallowed conservation laws of physics, the conservation of the number of heavy particles called baryons (neutrons, protons, etc.), can be violated; in the conditions of the early universe, quarks turn into leptons, and vice-versa. Such processes would be important only at times on the order of 10^{-35} sec after the big bang, when temperatures were 10^{28} K! This offers a possible way of understanding why our universe seems to have so few anti-baryons compared to the number of "normal" ones. It also predicts that the proton should be unstable, with a lifetime on the order of 10^{31-33} years. If the proton lifetime were as low as 10^{16} years, a million times the age of the universe, every human being would receive a lethal dose of radiation from the protons decaying in his body. If the universe is open, then after about 10^{32} years there would be no "matter" left; the universe would be a sea of neutrinos!

Of these attempts to unify the nongravitational forces of nature, Murray Gell-Mann, who in a sense founded the present approach to particle symmetries, has said, "I expect that there will be an intellectual revolution comparable to those that have taken place in the past with the heliocentric idea, evolution, special relativity, quantum mechanics." [33]

What we have described is only the first step in the revolution of which Gell-Mann speaks; but it already represents a far deeper understanding of nature than could have been foreseen only a few years ago. And we may yet find a way to unify all of the forces of nature, gravitational included, into a global unified theory. Says D. Z. Freedman of these theories of "supergravity," "a symmetry that unifies particles of different spins must be related to the structure of space-time, and therefore to gravitation. That's very appealing. The basic idea—that there can be a symmetry between bosons and fermions, [34] and that symmetry must be connected to space-time and that it is related to the force of gravity—must be right at some level." Gell-Mann says, "Some days we feel optimistic because we speculate intelligently about unification. Other days we feel less optimistic because we are dealing with such a huge extrapolation in energy beyond what we know. Most likely there are new phenomena in between that cloud the picture." [35]

Where this story of an intimate connection between the world on the smallest scale and on the largest will lead, one can only guess. But how remarkable it is, that we *know* what the universe was like only a few seconds after it began (see fig. 14.13)!

CHAPTER 15

Gravitational Collapse

No more revolutionary views of man and the universe has one ever
been driven to consider seriously than those that come out of pondering
the paradox of gravitational collapse, greatest crisis of physics of all time.

J. A. Wheeler[1]

"Overhead, without any fuss, the stars were going out"
Arthur C. Clarke, *The Nine Billion Names of God*

The picture of the structure of the universe on the grand scale we have
just completed is in a way unrealistic. The assumption of large-scale
homogeneity embodied in the cosmological principle dismisses as irrelevant to cosmology local or small-scale "clumpiness." Although we in-
habit a mediocre planet about a not very distinguished sun, one among
200 billion in our galaxy alone, to us the stars loom important. They
are the abode of life, of our own and of whatever civilizations may
share the vastness of the universe with us. But on the large scale, the
stars recede into insignificance; even the galaxies become but points,
the atoms, or elementary particles, of the universe. The scale of the
universe is 100,000 times that of a galaxy, the same ratio as the size of
an atom to its nucleus. For most purposes cosmology smooths out the
stars and galaxies and all other graininess and local irregularity into a
continuum. And yet we cannot totally ignore these tiny and dim points
of light, even when our focus is strictly on cosmological ideas. For it
turns out that the origin of the universe in a big-bang event and the
collapse of a massive star as it dies are processes that are the reverse of
one another. The equations of general relativity, which provide us with
a description of the beginning and the end of the universe, constitute
the language in terms of which the collapse of a star toward a space-
time singularity must also be given. In both cases it is the singularity,
this stage of finite mass contained within zero volume, that threatens to
make our physics and mathematics meaningless.

Whether nature is more tolerant of true singularities is an open question; certainly they exist, for the present, only in the mathematics. But the fall toward such a singularity in supernova collapse to a black-hole state requires very much the same description as the expansion from the singular state at the "origin" of the universe in the big-bang cosmologies. In the case of supernova collapse, gravitation overwhelms all other forces of nature in "squeezing a star out of existence," whereas in the primeval fireball, quantum gravitation initially dominates and then the other forces of nature hold sway until $t = 10^{12}$ secs, when matter becomes essentially neutral, after which time gravity again dominates.

So it is the phenomenon of gravitational collapse that we address here, focusing our attention on the collapsed state that may arise when a star several times more massive than the sun reaches the end of its evolution and undergoes a catastrophically violent death. The resulting black hole is the ultimate state of all matter if sufficiently compressed. Because this state of collapse from which no rescue is possible, seems to be most easily achieved in stellar collapse, we now set about to understand how stars come to this violent end.

Stellar Structure and Evolution

Although some understanding of the physics of stars was possible in the late eighteenth century, no important progress could be made in the study of the evolution of stars until the advent of quantum mechanics, the physics of the microscopic world—and its progeny, nuclear physics. Nineteenth century attempts to explain the enormous energy production of the sun as a result of ordinary chemical reactions (combustion) were doomed to failure, since "burning" the entire sun in this way could supply its present power output for only a few thousand years. This was no problem in the early eighteenth century, when the age of the earth was assumed to be only 6,000 years. But as the maturing sciences of geology and biology soon began to show,[2] the age of the earth, and therefore of the sun, is far greater than that.

The release of gravitational energy through slow contraction could furnish the required energy for some few million years, but again this is still far short of the known age of the sun. The knowledge and tools necessary to solve this problem had been accumulated by the late 1930s, when Bethe, Critchfield, and von Weizsäcker[3] were able to seriously propose that energy is generated in the interior of the sun through nuclear reactions.

Another crucial ingredient in our understanding of the nature of stars and of their structure and composition came from rapid evolution in the tools and techniques that have become available to the astronomer. Before the advent of photography and spectroscopy in the last century, all astronomy was visual, and hence mostly qualitative or descriptive. Spectroscopy, through analysis of the light from a star, made it possible for the first time to study the chemical composition of stars. As early as 1815 Fraunhofer had carefully measured hundreds of solar spectral lines and by 1872 Draper had analyzed hydrogen lines in the spectrum of the star Vega. But until World War II astronomy meant *optical* astronomy. Ever larger telescopes were turned toward the heavens, penetrating ever deeper into space. The cosmic distance scale was established, though it was to be modified several times. In the late 1930s and early 1940s, Karl Jansky at Bell Telephone Laboratories and Grote Reber in his backyard in Illinois "invented" the radio telescope, the tool that was to revolutionize astronomy in the latter half of the twentieth century. By 1960 radio astronomy had become an equal partner with optical astronomy in the search to uncover the secrets of the universe, and in only another 10 years astronomers had claimed almost the entire electromagnetic spectrum, by adding infrared, ultraviolet, x-ray, and gamma-ray astronomy to their repertoire. These developments were made possible mainly by rockets and earth satellites capable of transporting instruments above the earth's atmosphere. What is important about this is that at visible wavelengths, most of what we see is nearby and rather ordinary, while many of the brightest radio sources are at enormous distances, and the x-ray and gamma-ray objects are all exotic and violent.

Even without a theory of how energy is produced in the sun, the laws of classical physics tell us what conditions must be like in its interior and how energy is transported to the surface where it is radiated away. We find that in the center of the sun the temperature must be 10 million degrees or more and the density about 100 times that of water. Even at that density, the matter is in a gaseous state—or rather, since at such high temperatures electrons have been stripped away leaving charged nuclei and a "gas" of electrons, a "plasma" state. At these enormous temperatures, the nuclei collide with such energy that nuclear reactions take place.

The sun is composed mostly of hydrogen (about 75 percent), with the remainder mostly helium (nearly 25 percent); the heavier elements, including carbon, nitrogen, oxygen, magnesium, silicon, iron, calcium, which are so important as constituents of the earth and biological matter, total not much more than 1 of every 100 atoms in the sun. Fortunately for us the sun is a second- or third-generation star, having condensed from an interstellar medium enriched by the death of older stars that formed when the universe consisted almost entirely of hydrogen and helium. It is this fact that has led to the formation of small, dense, earth-like ("terrestrial") planets near the sun, with a high proportion of silicate and carbonate rocks, iron, and so on. Relatively young stars like the sun, "rich" in heavy elements synthesized in the interiors of earlier stars, are designated Population I, while the primordial stars, devoid of heavy elements, belong to Population II. Nucleosynthesis, the formation of the chemical elements, has occurred in the interior of stars, and, with the exception of hydrogen, the atoms in our bodies were long ago manufactured in the center of some now-dead star!

Our knowledge of the conditions of equilibrium in stars—the balance between the gravitational forces tending to make a star contract and the outward pressure that results from the high central temperature—coupled with the development of the quantum theory in the 1920s and nuclear physics in the 1930s, led to a model of energy generation in stars that is now generally accepted.[4] At the high temperatures prevailing in the center of a star, some 13 million degrees K in the case of the sun and much higher in more massive stars, hydrogen is converted into helium through nuclear reactions, mainly the proton-proton cycle in stars like the sun and the carbon-nitrogen-oxygen or C-N-O cycle in heavier stars (fig. 15.1). The enormous amount of energy thus released is transported from the center to the surface of the star, from which it is radiated as heat and light.

In this stage of its life, a star is said to be a "main sequence" star because, during this stable hydrogen-burning phase, representing about 90 percent of its useful "life," the brightness of a star, as well as its color and temperature, is directly correlated to its mass: very massive stars are hot and blue-white to blue; very small stars are cool and red. These relationships are most readily exhibited in the "Hertzsprung-Russell diagram," shown in figure 15.2, in which the brightness (luminosity or power output) is plotted against temperature. The main sequence is seen as a band extending diagonally across the diagram, from cool, red stars ("red dwarfs") in the lower right hand corner, to hot, blue stars ("blue giants") in the upper left. The sun, an ordinary main sequence star, yellow-white to white in color, has a surface temperature of 6000 K and occupies (as it has for nearly 5 billion years) a more or less central position in the diagram. Because the brightness of a star, and hence the rate at which it consumes its nuclear fuel, depends on the third or fourth power of its mass,[5] the more massive the star, the shorter its lifetime (approximately inversely proportional to the square of the mass). Stars 10 times more massive than the sun may proton–proton (p–p) cycle:

$$^1H + {}^1H \longrightarrow {}^2H + e^+ + v$$

$$^1H + {}^2H \longrightarrow {}^3He + \gamma$$

$$^3He + {}^3He \longrightarrow {}^4He + 2\,{}^1H$$

C–N–O (carbon) cycle:

$$^{12}C + {}^1H \longrightarrow {}^{13}N + \gamma$$

$$^{13}N \longrightarrow {}^{13}C + e^+ + v$$

$$^{13}C + {}^1H \longrightarrow {}^{14}N + \gamma$$

$$^{14}N + {}^1H \longrightarrow {}^{15}O + \gamma$$

$$^{15}O \longrightarrow {}^{15}N + e^+ + v$$

$$^{15}N + {}^1H \longrightarrow {}^{12}C + {}^4He + \gamma$$

e^+ = positron

γ = gamma ray

v = neutrino

Figure 15.1 Hydrogen-burning nuclear cycles in stars. The proton-proton cycle is the dominant nuclear source for the sun. In the first step two protons (hydrogen nuclei, ^1H) combine to produce a deuterium (heavy hydrogen, ^2H) nucleus plus a positron (antielectron) and a neutrino. The neutrino immediately escapes from the sun. In the second step the deuterium nucleus combines with a proton to form helium-3(^3He), a light isotope of helium that differs from the heaviest isotope of hydrogen called tritium, only by having neutrons and protons interchanged. A gamma ray is also emitted. In the third step two helium-3 nuclei combine to yield one helium-4 (^4He) nucleus, but two protons are emitted as well. Simple counting will show that the net result of the cycle is that four protons fuse to form one helium-4 nucleus. There is a mass loss of nearly 1%, which is released as energy (of the particles involved). The carbon or C-N-O cycle dominates in more massive stars than the sun, where the central temperature is considerably higher (C = carbon, N = nitrogen, O = oxygen).

reside on the main sequence for only 100 million years ($^1/_{100}$th the time of the sun). Stars more massive still will have even shorter lifetimes, while red dwarf stars, as much as 10 times less massive than the sun, may live for a trillion years.[6]

After the long hydrogen-burning phase, a star will begin to move off the main sequence, becoming cooler and redder and, in the case of modest stars like the sun, becoming brighter. This can happen only if the star begins to swell,[7] a condition that occurs as it begins to run out of central hydrogen; its core contracts and heats up to the point at which

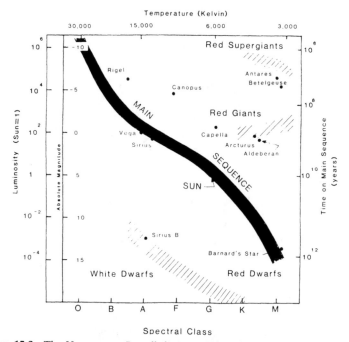

Figure 15.2   The Hertzsprung-Russell diagram. When the brightness (luminosity) of a collection of stars is plotted versus their color or temperature or spectral class (O-B-A-F-G-K-M, from hot to cool stars), about 90% of the stars fall on a diagonal band known as the main sequence, while most other stars occupy a small number of other significant areas in the diagram. The most massive stars spend only a few million years on the main sequence, while the least massive may remain there for a trillion years.

helium can begin to "burn" through the "triple-alpha" process, in which three helium nuclei fuse into one carbon nucleus ($3He^4 \rightarrow C^{12}$), and its envelope, or atmosphere, balloons to an enormous size. For the sun, this will be its "red-giant" stage, when it will be perhaps 50 times its present size, much cooler, but as much as 100 times brighter because of its large surface area (fig. 15.3). Temperatures on the earth will approach 1000 K, and, as the oceans and the atmosphere begin to evaporate, life will become impossible on its surface. Following this relatively brief helium-burning epoch, the sun will slowly collapse toward its ultimate "death" as a white dwarf star.

For more massive stars, the scenario will be more complex and more violent. Not only will helium burning take place, but also carbon burning, oxygen burning, etc., up to the iron group of elements where the thermonuclear fusion reactions we have been describing will no longer release energy but instead will begin to absorb it.[8] The star will have

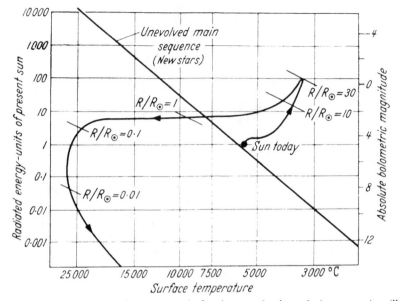

Figure 15.3 Presumed evolutionary track for the sun. In the red-giant stage it will be nearly 100 times its present brightness, and temperatures on earth will rise toward 1000 K. Subsequently the sun will contract to become a planet-sized white dwarf, hundreds, then thousands of times fainter than at present. Temperatures on the earth will be only a few degrees K. G. R. Burbidge and E. M. Burbidge, "Stellar Evolution," Handbuch der Physik, vol. 51, Springer-Verlag, 1958.

become a red supergiant (examples are Betelgeuse and Antares) perhaps 25,000 times brighter than the sun and will have swollen to 500–1,000 times the size of the sun. At this stage it is confronted by a truly violent death: radiating enormous amounts of energy, but devoid of nuclear fuel, it collapses catastrophically, giving rise to a supernova explosion.[9]

Supernova Collapse

A supernova event is the most violent transient phenomenon we know and involves for all practical purposes the destruction of a star. Some significant fraction of its mass will be ejected into space, its core will be compressed to almost inconceivable densities, and the star will, for a time, increase in brightness by more than a million times. Such outbursts in our galaxy have been recorded as "new stars" on several occasions, including the one in July 1054, which was observed by Chinese astronomers and probably American Indians[10] and which gave rise to

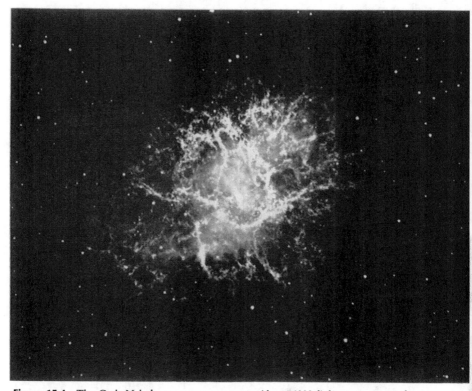

Figure 15.4 The Crab Nebula supernova remnant. About 6000 light years away, the Crab supernova erupted in the constellation of Taurus in July 1054. Near the center is the neutron-star pulsar NP 0532, which pulses every 0.033 seconds. This means that its radius cannot exceed 700 km (see the text) and in fact is probably only about 10 km. Palomar Observatory Photograph.

the Crab Nebula (figs. 15.4 and 15.5). That it was "missed" by Arab astronomers is one of the great mysteries in the history of astronomy, possibly a tribute to the strong hold that the Aristotelian image of the immutable heavens had on the medieval imagination, both Latin and Arab. It is difficult to escape the conclusion that this new star, perhaps bright enough to have been seen in broad daylight and visible to the naked eye for nearly two years, appearing in the morning sky just before dawn and dominating the fall sky for months, was not recorded in medieval chronicles, because, according to Aristotle, it could not have been a celestial object! Only recently has a medieval account of this even been found, dramatic testimony to the power of a paradigm. We are now able to identify this ancient supernova with the Crab Nebula, an irregular cloud of expanding gas located about 6000 light-years from

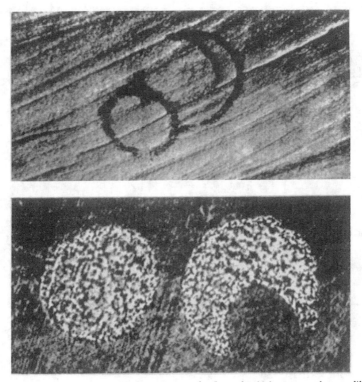

Figure 15.5   American Anazazi Indian pictographs from the 11th century that are likely to be representations of the 1054 Crab supernova outburst. William C. Miller, Hale Observatories.

the earth. By knowing its size (about 3 light-years) and by measuring the rate of expansion of the nebulosity, it can be shown that the explosion must have occurred about 900 years ago. There are a number of extraordinary aspects to the Crab Nebula, some of the more intriguing of which will be touched upon later in this chapter.

There have been several other "new stars" recorded in our own galaxy in historical times that can be more or less definitely ascribed to supernova outbursts, among them were those which occurred in the years A.D. 185 and A.D. 393.[11] In addition, there was a well-documented case of an especially brilliant "guest star" in the constellation Lupus in 1006, seen at Swiss and Italian monasteries and by Oriental and Arab astronomers; there was another in 1181 in Cassiopeia. The best known are Tycho's nova of 1572, also in Cassiopeia, whose role in shattering the Aristotelian "two-sphere universe" has already been discussed, and "Kepler's nova" of 1604, in Ophiuchus. Another new or "guest" star observed in the fall of 1408 by Chinese astronomers may well have been the supernova outburst that produced Cygnus X-1, which is at present our best candidate for a black hole (see below). Remnants of these outbursts are difficult to detect, and we are able to locate them mainly through their x-ray and radio emissions. Since supernova outbursts are observed in other galaxies at the rate of about 1 every 50 years per galaxy, we might be thought "overdue" for a supernova in our galaxy. But interstellar absorption by galactic dust and gas would make a supernova at a typical distance of 30,000 light-years just at the limit of naked-eye visibility and therefore likely to be missed.

The energy output of a supernova is enormous, typically a billion times that of the sun. Such a star would be much brighter than the full moon if at a distance of 32 light-years from the sun (10 parsecs) and would be nearly as bright as the sun if it were as close as α Centauri, the nearest stellar system to our own. In several cases supernovae have shone with such brightness that their light output has exceeded that of the entire galaxy in which they were situated, and in one case, that of a supernova discovered in the galaxy IC 4182 in 1937, the exploding star was 100 times brighter than the rest of the galaxy!

The mechanism by which the energy released in supernova collapse blows away the outer layers of the star is not fully understood, but one explanation is that the enormous numbers of neutrinos generated at the high temperatures in the interior of the collapsing star deposit their energy in its envelope carrying it away. Modern gauge theories of weak interactions offer a hope of full understanding of this process (see chapter 14). Another possibility is that a shock wave resulting from the implosion propagates outward through the star, blowing away its outer layers.

The final result of this implosion-explosion process is a collapsed remnant and an expanding shell of gas such as we see in the Crab Nebula. The nature of the core will depend on the mass it contains after the envelope is blown away. If a large fraction of the mass is ejected or converted into energy, the core might stabilize as a white dwarf star, with a mass less than about 1.5 solar masses, a limit derived by the Indian astrophysicist S. Chandrasekhar in 1931. This is a more likely outcome of the death of ordinary stars like the sun, the "stellar bourgeoisie." A neutron star will result if the final mass is only a little greater, about two solar masses. But if the star should fail to shed enough matter to reach this limit, it must continue to collapse forever toward a singularity, becoming a "black hole."[12] It is this pathological state that we wish to concentrate on, because of its cosmological implications, but first a brief survey of the properties of the other "stellar corpses," the white dwarf and the neutron star, is in order.

White Dwarf and Neutron Stars

A white dwarf star is a collapsed star whose contraction has been halted by "electron degeneracy pressure," a quantum mechanical effect that is a consequence of the "Pauli exclusion principle." In such a star the electrons constitute what is called a degenerate electron gas, and such a state of matter strongly resists further contraction. Only one electron can occupy a given state, and when a configuration is reached such that all the lowest states (essentially energy levels) are filled, no further loss of energy is possible for the electrons. The nuclei of the atoms can continue to lose energy so that the star radiates, albeit weakly, gradually becoming cooler. The star will have shrunk to the size of a planet only a few thousand miles in diameter. Because the electrons will efficiently conduct heat, the white dwarf star will be nearly the same temperature throughout. Its surface is relatively hot, which makes it blue-white or white, if young, but its interior is cool as stars go, not the several millions of degrees typical of a normal star. Because the star has shrunk by a factor of 100, its density will have increased by 1 million times, to something like 10^6 g/cm^3. A teaspoonful would weigh a ton at the earth's surface, and 10,000 tons at the surface of the star. The most famous white dwarf star is "Sirius B," the companion of Sirius. A hot blue-white star nearly 500 times less luminous than the sun, it is only about 10,000 miles in diameter, even though its mass is nearly identical to the sun's. An interesting fact about white dwarf stars is that the more massive they are, the smaller (in diameter) they are.[13]

As we have seen, the result of supernova collapse, if enough mass is shed, could be a neutron star. This comes about when the density of the collapsing star becomes so great that protons and electrons (through the weak interaction) combine to produce neutrons: $p + e \rightarrow n + v$. When this happens, normal atoms disappear because they lack the protons that define them chemically, and the electrons that previously halted the collapse of the white dwarf star are gone, so that continued collapse to nuclear densities is possible. The result is a star composed mostly of neutrons—nearly pure neutron matter—with a diameter of but a few miles and a density of about 100 million times that of a white dwarf.

Such a star would probably be rapidly rotating as well. The sun rotates once in about 26 days, but if it were to shrink to neutron star size, it would spin many times a second.[14] This rapid rotation, coupled with the incredibly intense magnetic fields that result from the collapse, apparently give rise, we believe, to what is known as a "pulsar." Pulsars are thought to be rotating neutron stars emitting searchlight-like beams of radio energy, x-rays, etc., which sweep past the earth at the rate of once every few seconds to a fraction of a second (fig. 15.6). The most rapid pulsar known is the Crab supernova remnant, a neutron star rotating 30 times per second! The Crab pulsar is young; older pulsars have longer periods, having slowed down as energy has been extracted from their rotation.

In late 1967 radio astronomers at Cambridge University discovered a source of radio signals that pulsed at a rate of precisely 0.7477747 times per second. The knowledge that no known natural source could emit such accurately timed radio pulses led to its being dubbed facetiously, if somewhat hopefully, "LGM," an acronym for "Little Green Men." It was soon obvious that the pulses were not the product of intelligent life elsewhere in the galaxy; mitigating against this interpretation, for example, was the lack of any planetary motion on the part of the pulsar source. Evidently, the source was stationary and was presumably a star the size of a white dwarf or smaller.[15] Thus, the LGM met its demise and the "pulsar" took its place. By now more than 300 pulsars are known with periods ranging from 0.03 to 4 seconds. These periods,

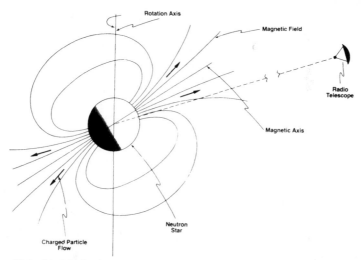

Figure 15.6 Model of pulsar emission. The axis of the magnetic field of the neutron star is inclined with respect to the rotation axis. The beam of synchrotron radiation sweeps past the earth, causing the characteristic pulse at radio, x-ray, and even visible wavelengths. "Stellar Ontogeny . . . to Ashes," E. Edelson, *Mosaic,* May/June 1978, National Science Foundation, 1979.

while known with great precision, have been observed to be very slowly increasing, indicating a loss of energy and angular momentum by the pulsar to its surroundings. The mechanism by which a rotating neutron star might emit pulsed radio waves (and in some cases x-rays and even visible light) is not fully understood, but the most likely suggestion is as follows. It is thought that the magnetic and rotational axes of the rapidly spinning star are inclined with respect to one another. Electrons emitted at the magnetic poles spiral about the magnetic lines of force, emitting electromagnetic radiation called "synchrotron radiation," which, as the star rotates, sweeps by the earth. The result is that we see a very short pulse of electromagnetic radiation. The magnetic field possessed by a neutron star is so intense (on the order of 10^{12} gauss, compared to only a few gauss for the sun) that it can be thought of as almost rigid, fixed in the crust (see below) of the star.

The rate at which the rotation of the pulsars is slowing is consistent with their interpretation as rotating neutron stars. The Crab nebula pulsar, NP 0532, with a period of 0.033 sec, is slowing at the rate of about 4×10^{-6} sec per year, corresponding to an energy loss at a rate of 10^{38} ergs/sec or 10^{28} kw.[16] Furthermore, their periods have been observed to suddenly undergo small discontinuous changes, followed by a slow return to normal after a period of weeks or months. These changes have been interpreted as "starquakes," shifts in what appears to be a rigid surface or crust of the star! The final picture that emerges is of a rapidly rotating neutron star, the source of an extraordinarily intense magnetic field, with a rigid crystalline crust, probably composed of neutron-rich heavy elements, with an interior of superfluid neutrons. The coupling between the crust and interior is weak, so that when a starquake occurs, it takes perhaps as much as a year before the crust and core again rotate as a single rigid body.

Black Holes

The ultimate in collapsed stellar configurations, a star whose shrinking cannot be halted under any circumstances (if we believe general relativity), is the *black hole*. When the mass of the collapsed core exceeds two or three times the mass of the sun, the collapse will continue without limit. In what follows, we shall examine in some detail the properties of these peculiar objects, an understanding of which is impossible without general relativity.

Recall that in the late eighteenth century Laplace argued that if a star were compressed to the radius $2GM/c^2$, the escape velocity from its surface would equal the velocity of light. The modern theory of collapsed stellar matter was developed by Oppenheimer and Snyder in 1939.

One of the consequences of this work is that, when a contracting star falls within a certain radius, now called the Schwarzschild radius,[17] light emitted from its surface is red shifted to zero energy. That is, the energy expended in climbing up out of the intense gravitational field of the star equals the total energy of the proton. No light can escape, nor can any other radiation. This critical radius is also known as the "event horizon," and it is a "one-way membrane," in that, although matter and radiation can fall in, nothing can escape. In a sense, the star has left the universe, leaving behind only its gravitational field, which cannot "know" what has happened to the star. Such an object is a black hole, and at its center is a space-time singularity, where the density becomes infinite. To an observer riding on the surface of the collapsing star, such an infall would take only a short time, perhaps a few seconds; to a distant observer (us) the time for collapse to the Schwarzschild radius $2GM/c^2$ would be infinite. This is because a clock carried by the observer falling in with the star would be retarded by the intense gravitational field (an aspect of the gravitational red shift), and at the event horizon his clock would stop. A distant observer, however, would see the star getting dimmer and dimmer, redder and redder. Its light would shift into the infrared, then to radio wavelengths and to lower and lower energy. Its energy output would become more and more feeble. Very quickly, the star would become undetectable. J. A. Wheeler's description of this extraordinary object is worth quoting:

What would be the appearance of the collapsing core if it could be seen from afar without the interference of the supervening envelope? The hot core material is brilliant and at first it shines strongly into the telescope of the observer. However, by reason of its faster and faster infall it moves away from the observer more and more rapidly. The light is shifted to the red. It becomes dimmer millisecond by millisecond, and in less than a second too dark to see. What was once the core of a star is no longer visible. The core like the Cheshire cat fades from view. One leaves behind only its grin, the other, only its gravitational attraction. Gravitational attraction, yes, light, no.

No more than light do any particles emerge. Moreover, light and particles incident from outside emerge and go down the black hole only to add to its mass and increase its gravitational attraction. Has the black hole a size? In one way, yes; in another way, no. There is nothing to look at. One could of course imagine thrusting a meter stick toward the center of attraction until it "touched base." However the powerful tidal forces will tear apart that object and every other object. No conventional measurement of the dimensions is possible. Even to speak about the dimensions of the object is any conventional sense is out of the question.[18]

What is left, then, is an indefinitely collapsing core, the source of an incredibly intense gravitational field at small distances but totally invisible, emitting no radiation of any kind. This singularity[19] in space-time can be detected only by its gravitational field, which would "swallow" any passing object, including a light ray that might be directed sufficiently near it. Any massive object, such as an unlucky spacecraft, passing very near, would be torn apart by the immense tidal forces created.

If general relativity is correct, nothing can stop the collapse of a massive star toward a singularity; general theorems have been proved by Hawking and Penrose[20] and others that establish this fact. The Schwarzschild radius for the earth is about 1 cm, while for the sun it is about 3 km. If the sun were to collapse or be compressed to a radius less than this, it would continue to collapse forever (as seen by a distant observer) and fade from view as a black hole.

Figure 15.7 represents the fall of an observer into a black hole, in what are called "Kruskal"[21] coordinates. After crossing the "event horizon," no signal sent back toward the outside could escape; it would be captured by the singularity. Similarly, no rocketship, no matter what the power of its engines, could even remain stationary above the singularity; it must fall inexorably to the center. An observer falling into the black hole would reach the singularity only milliseconds after crossing the event horizon. In fact our observer would have long since been turned into "spaghetti" by the enormous difference between the gravitational force at the top of his head and the tip of his toes!

As we have already seen, the remnant formed in supernova collapse, whether a neutron star or black hole, will likely be rotating rapidly, and the case of a rotating black hole is different and more complex than the description already given. The singularity is a ring, rather than a point, so that a rotating black hole has a "throat," and the collapse does not continue all the way to zero radius. The consequence of this is what Einstein called a "worm hole," sometimes known as an "Einstein-Rosen bridge." In such a case, which was first described by the British astrophysicist Roy Kerr, one can imagine paths into the black hole that miss the singularity and could emerge "elsewhere" or "elsewhen"!

This has led to "cocktail-party" speculation that one could travel through a black hole to another universe (an idea that has no operational meaning—no information could be returned about such an experience; it is incapable of empirical verification) or to another part of our universe.[22] But the value of such speculation is dubious, especially since it ignores the presence of the matter of the star within the event horizon—the Kerr and Schwarzschild solutions are valid only where there is no mass.

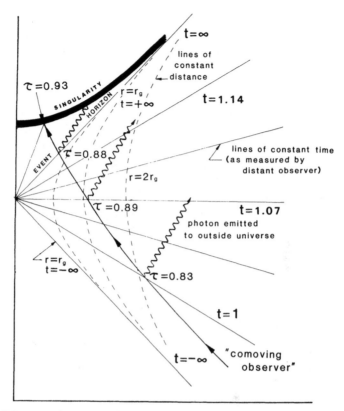

Figure 15.7 Kruskal diagram showing the free fall of an observer into a black hole. The observer might be thought of as sitting on the surface of the collapsing star. Photons emitted after passage through the event horizon ($r = r_g = 2\ GM/c^2$) cannot escape; they hit the singularity. The times labeled by τ are those read by the observer's clock as he falls in. Those labeled by t are those of a distant observer.

A crucial feature of a black hole is that it can be characterized by only its mass, its angular momentum (its spin), and its charge, if it is not neutral. All other information is lost. John Wheeler has described this fact by saying "a black hole has no hair," by which he means that black holes have no distinguishing characteristics, no individual identity—except their charge, mass, and spin. Some of the most revered conservation laws of physics are violated in the interior of a black hole.

How then could a black hole be observed? We could imagine detecting one by means of the gravitational "lens effect," due to the effect of the gravitational field of the black hole on light passing near it, but it is difficult to see how this could be carried out in practice.

Unquestionably the most promising approach involves the search for components of close binary star systems that may be black holes. A double star system consisting of a normal star and an invisible companion of very large mass is a good candidate for a black hole. In such a case, if the stars were sufficiently close to each other,[23] matter would stream from the normal star into the black hole, causing the emission of x-rays (fig. 15.8). Thus, a double star system involving 10–30 solar masses, in which one component is invisible, and which is, in addition, a source of x-rays with the correct signature (principally the rate of flickering), would be an excellent prospect. A few such objects have been found, and one, Cygnus X-1,[24] may be the first black hole actually detected (fig.15.9).

By their very nature difficult to detect, black holes remain the creation of the theoretical astrophysicist, with no unequivocal evidence of their existence. On the other hand there is little reason to doubt that they exist, and as we remarked earlier, it is possible that a significant fraction of the mass of the universe is bound up in black holes. Whether this is indeed true remains to be seen, but it may be that first-generation stars, those formed in the early days of the universe, were extremely massive and thus likely to end up as black holes. Another interesting possibility is that huge black holes are at the center of galaxies, including our own. Many galaxies have intensely active nuclei, and some, such as Seyfert and "N" galaxies, are strong sources of radio energy. The nucleus of our galaxy is less active, but at infrared and x-ray wavelengths, with which we can see to the center, the nucleus is extremely bright, and has a temperature of 10 billion degrees; 25 percent of the radiation comes from a region smaller than 10 astronomical units (a.u.'s) in diameter (the size of Jupiter's orbit), strongly suggesting a collapsed object. In giant elliptical galaxies the mass-to-light ratio is very large,

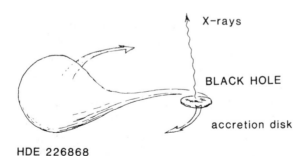

Figure 15.8 Model of Cygnus X-1, presumed to consist of a black hole and a normal star, orbiting a common center of mass.

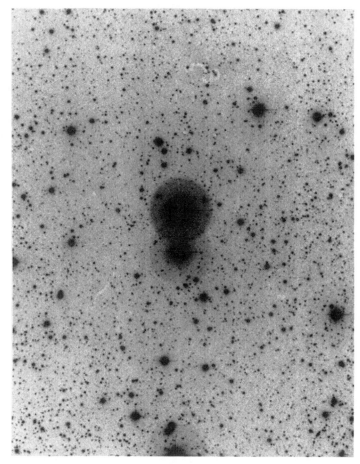

Figure 15.9 Cygnus X-1. The star HDE 226868, a spectroscopic binary star thought to contain a black hole. Palomar Observatory Photograph.

indicating the presence of a large amount of nonluminous matter, perhaps in the form of central massive black holes. Furthermore, the quasars may be violent events in the nuclei of galaxies caused by the collapse of one hundred million stellar masses of material to massive black holes.

Further speculation about gravitational collapse and the density of matter in the universe has led to the suggestion that the entire universe is on the verge of collapse toward a black hole state. On the basis of a reasonable assumption about the amount of matter in the universe, its Schwarzschild radius should be on the order of a billion light-years, perhaps more, depending on its density (chapter 12). This figure is of the same order of magnitude as one might give for the present radius of the universe (say 18 billion light-years). The difficulty is that, while the description of black holes (the Schwarzschild solution) has to be "patched" onto the rest of the universe, the black hole has an outside; the space-time geometry of a black hole is flat at large distances and has to merge with the rest of the universe. The universe, of course, has no outside.[25]

It is interesting that, although a black hole results from the collapse of a massive object beyond the critical Schwarzschild radius, the critical density is not necessarily immense or even large. When a star the mass of the sun or of several suns collapses into a singularity, the densities approach and exceed that of nuclear matter, but if the object is much larger the picture is quite different.[26] The greater the mass, the lower the density at which an object becomes a black hole. Figure 15.10, in which mass is plotted versus radius, reflects this result. The present average density of the sun is about 1.3 g/cm³, slightly greater than water, but as we have noted, if it were compressed to the Schwarzschild ra-

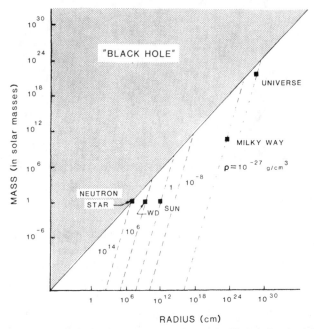

Figure 15.10 The relation between the "radius" of a black hole (the Schwarzschild radius $2\,GM/c^2$) and the mass of an object in solar masses. The corresponding densities are shown. One notes that the greater the mass, the lower the required density (which is inversely proportional to the mass squared).

dius, about 3 km, its density would be greater than 10^{15} g/cm^3. On the other hand, the Schwarzschild radius for the Milky Way galaxy, with a mass of 200 billion solar masses,[27] is about 3×10^{16} cm, or 5,000 a.u., a dimension comparable to that of the solar system with its halo of cometary debris. But the density of the collapsed galaxy would be only 10^{-8} g/cm^3 as it became a black hole, a figure 100,000 times less than the density of air.

Similarly, the tidal forces experienced by an object thrust into a black hole of stellar mass would be immense, sufficient to tear apart any material object. But the gravitational field gradient (rate at which the field changes with distance) is *inversely* proportional to M^3 so that at the "edge" of a black hole of the mass of a galaxy, the field gradient would be 10^{33} times gentler than in the case of a collapsed star. One could walk into such a black hole and hardly notice anything strange (at least as far as gravitational effects are concerned).

At the other end of the distance scale are the "mini black holes" proposed by Stephen Hawking.[28] The inclusion of quantum mechanics in the description of a black hole seems to permit them to "evaporate" through a quantum mechanical "tunneling" process; they radiate particles and photons, ultimately vanishing! Hawking suggests that such objects might have been produced in large numbers in the first second or so of the big bang, and while the smaller ones would have disappeared by now, larger ones would still be radiating and would be seen as x-ray sources. This thermal radiation from an event horizon is called "Hawking radiation."

The death of a star, then, may be violent or slow and unremitting. In the latter case, the star, a black or white dwarf, will fade slowly into obscurity, eventually becoming a nameless cinder in the blackness of space. In the former, however, death leads to rebirth. For the debris cast into interstellar space is the stuff out of which new stars are born. The original matter of the universe, mostly hydrogen, is constantly being processed in the interior of stars, enriched through the synthesis of heavy elements, then cast again into the vast space between the stars, there to form the raw material out of which new stars will condense. The picture is one of birth, death, rebirth, inexorably constrained by the law of entropy.[29]

Epilogue

> The effort to understand the universe is one of the very few things that
> lifts human life a little above the level of farce and gives it some of the
> grace of tragedy.
>
> Steven Weinberg

"A Simple and Intelligible Picture"

As much as at any time since Galileo first picked up a telescope,
cosmology is in an observational phase. As the data from orbiting telescopes, long-baseline radio arrays, and low-background counting experiments accumulate, some of the fundamental questions about the
structure of the universe will be decided. On the theoretical side, the
possibility of surprising developments is equally great. The already
substantial implications for cosmology of elementary particle physics
and "grand unified theories" can only multiply, and there is hope that
the much-sought accommodation of gravitation to quantum physics will
be achieved.

Is this period of excitement and innovation to be primarily one of
consolidation, like that which saw Newtonian dynamics applied to the
planets? Or are some of the puzzling or obscure observations truly
anomalous, as the ether experiments of the nineteenth century proved
to be? It will be fascinating enough if the new instruments can provide
a clear verdict on whether the universe is finite or infinite. The cultural
impact of an unambiguous "measurement" of the universe could be
substantial, though the public seems willing enough to ignore the question as long as no resolution is forthcoming. Some astronomers and
physicists have weighed in with opinions on this question, the open
universe lobby forming a small and undogmatic majority. But conservatism is the norm, as the record we have examined suggests it should

be. No one wants to be remembered with those unfortunate professors of a century ago who, misreading the status of electromagnetism, pronounced Newtonian physics complete and final, just when its crucial shortcomings had been found.

In sharp contrast to the unwillingness of scientists either to commit themselves prematurely to the details of a model or to contend that the basic physics itself could not be overthrown, is the attitude toward speculation within the bounds of broad general principles. Here the greatest latitude for speculation is allowed, indeed required. A particular gift for imaginative variations on the accepted themes is what brings Nobel Prizes. Later, when possibilities are narrowed by improved observations or a sufficiently compelling theory, conventionality is again encouraged and emotions damped. Physical cosmology is at a particularly interesting moment in this regard. The answers are tantalizingly near, while the range of serious speculations is broad. For our concern with the historical overview, it is also interesting that the subjective sources of many of the questions about the universe are still clearly visible.

In this connection the spirit of Einstein is strong. His emphasis on the elegant and beautiful will be influential after the particulars of general relativity are supplanted by whatever mathematical construct amalgamates quantum and field theoretic methods. In his address on Max Planck's sixtieth birthday in 1918, he expressed his position categorically.

Man tries to make for himself in a fashion that suits him best a simplified and intelligible picture of the world; he then tries to substitute this cosmos of his for the world of experience, and thus to overcome it.[1]

As many of Einstein's public statements are, this one is extraordinarily thoughtful and quite revealing. The repeated substitution of a new world picture for an earlier, less satisfactory one is essentially the narrative we have traced. Einstein speaks, however, not of one theory superseding another, but of a flight to a mathematical world, one characterized by "supreme purity, clarity and certainty," and capable of providing "the theory of every natural process, including life, by means of pure deduction, if that process of deduction were not far beyond the capacity of the human intellect." The researcher is motivated by a "longing to behold this preestablished harmony . . . akin to that of the religious worshiper or the lover." A high standard indeed, and one not likely to survive the transition to textbooks that is the last step for a successful science.

Limited Futures

Not every proposed cosmology can be taken seriously. It goes without saying that irrational or arbitrary world systems are ignored by
scientists, even if the systems are not falsifiable. "Divine Providence,"
which ascribes each action and event to the direct will of God, is ruled
out because, as Henry Morgenau observed, "It is too blunt; it lacks,
strange to say, the appropriate degree of metaphysical refinement."[2]
This and other religious systems "violate the rules of the game," because they make no predictions. The situation, however, is not as neat
as a division between the acceptably mathematical and the unacceptably
qualitative. Theories of the universe evoke strong emotional reactions,
as is evident in the question of whether the universe is open or closed.
The variety of reasons for rejecting mathematical cosmologies is illustrated by a look at an "unsuccessful" system.

In 1957 Hugh Everett, encouraged by the ingenious John Wheeler,
described a system for meshing traditional theories of matter with the
language of quantum theory.[3] A brief digression here will set the stage.
It has been clear since the 1920s that the basic product of a quantum
calculation is a set of probabilities for the occurrence of possible positions and velocities of the thing described. The "trajectory" of the object is prescribed by the way that the probabilities are altered by the
environment. For example, the environment might be a magnetic field,
the object a proton. The startling difference between the quantum
probabilities and those we are accustomed to deal with, such as the
probability of winning a raffle or of having a male offspring, is that the
quantum probabilities are not mutually exclusive. An object may have
a finite probability of being anywhere or of having any arbitrary velocity at a particular instant; in a *sense* the single electron of a hydrogen
atom fills the entire volume of the atom, for example. Nonetheless the
electron is never directly observed at two positions simultaneously. The
common interpretation, attributable to Max Born and Niels Bohr, is
that when the electron is subjected to precise measurement, the measured property—position or velocity—arbitrarily assumes one of its
possible values, the quantum probabilities determining the most likely
"collapsed" state. The stability of large objects such as this book is thus
statistical. The probabilities of remote positions or altered shapes are
small; the enormous number of electrons in the book multiply the un-
likelihood of discontinuous motion. It is not difficult to imagine, however, a realistic situation in which some large-scale future would be
determined by a single quantum transition. A single electron could, let
us say, trigger a large explosion. In this case the future "trajectory"

would not be predictable by any physical theory—or, of course, by any other means.

The interpretation of quantum measurement is widely accepted as we describe it, although Einstein broke with the quantum theorists over the discontinuity element, and many physicists agree that the arbitrary, statistical "choice" by electrons is unaesthetic. Everett's interpretation (called EWG, for Everett, Wheeler, and Neill Graham who extended it) eliminates the arbitrary collapse of the probability distribution, the feature that Einstein objected to. But the arbitrariness is altered in a most exotic way. EWG proposes that all the futures corresponding to all nonzero quantum probabilities actually occur in multiple, coexisting universes. The universe divides repeatedly into many different versions of itself, evolving in as many ways as are consistent with the laws of physics. The bomb in our example both explodes and does not explode. The book remains intact in most universes, but in a few it moves or disintegrates. To quote Bryce DeWitt, "Every quantum transition taking place in every star, in every galaxy, in every remote corner of the universe is splitting our local world into myriads of copies of itself."[4]

The reader would be in good company if his reaction to splitting universes were "that's just silly." But what are the objections? To the criticism that all those coexisting universes should reveal themselves, the advocates respond that the same argument was advanced to disprove the earth's motion. It is probably accurate to say that the typical scientist rejects EWG, if he knows of it, because the concept of multiple universes is uneconomical. Furthermore, the restriction of conscious observers to a single set of branches seems no less arbitrary than the quantum probability collapse. Whatever its difficulties, EWG serves to remind us that the universe is ultimately made up of quantum particles and that a fully satisfactory cosmology must deal explicitly with the relationship between this stratum of reality and large-scale theories such as general relativity.

Relatively few of the scientists who quickly dismiss EWG would be willing to follow Einstein in rejecting the claim of quantum theory to completeness. Yet such personal judgments are important in the early stages of the development of theoretical constructs, and more than usually so in cosmology. Thus, although the cosmological principle, expressing the spatial uniformity of the universe, is now strongly confirmed by the isotropy of the cosmic microwave background radiation, it was firmly entrenched for decades before the background radiation was discovered.

Gerald Holton has developed a rather formal language for discussing this "third dimension" of science, as he designates the preconceptions and biases that cut across theory and observation and are often more persistent than the theories themselves.[5] A persistent idea is called a "thema" (plural, "themata"). The conservation of energy, the second law of thermodynamics, and similar principles are not themselves themata in Holton's usage. Rather, such mathematically stated principles are undergirded by themata, which have the status of beliefs. Einstein's "longing for preestablished harmony" is an unusual, conscious reference to the Pythagorean thema of mathematical harmony, which is as manifest in the concern of modern physics for symmetry as it was in the Greek preoccupation with circularity-sphericity in the heavens.[6]

Other common elements of the Greek philosophies we discussed in chapters 4 and 5 can be designated as themata. The world as organism, the preference for teleological explanations, the macrocosmic-microcosmic correspondence, the continuum and its opposite atomicity—all have persisted, though not always in the mainstream of science. The mutually exclusive pair continuum-atomism (called "thema-antithema" by Holton) is interesting to trace since one or the other of the rivals has been ascendant in physical explanations since Democritus. The conceptual difficulty in quantum theory can be characterized as arising from the unwillingness of matter to conform to either discreteness or continuity.

The Once and Future World

The big bang cosmology seems at this point to be holding the field against its challengers. Because it incorporates a "creation event," it revives interest in what might be called "physical cosmogony." In Holton's terminology the antithetical pair stasis-creation can be labeled thema-antithema. The tension between natural philosophy and religion derives in no small part from their divergence on this issue. A clear line running from the Pythagoreans through Plato and Aristotle to Kepler, Newton, and mathematical science assigns to physical law a static character that would tie the hands of the Judeo-Christian-Muslim Creator. The expanding universe, though obviously not Einsten's first choice, is not necessarily a threat to the statis thema, because the particular initial conditions are only incidental to the field equations. The expansion can be traced from a state arbitrarily near the initial singularity—a state, however, with matter and time already existing. The remoteness of this state of unimaginable compactness might seem to bring it closer to a true creation event, but the relativistic cosmologies no more provide their own genesis than Newton's laws could produce point masses.

One aspect of the quantum description of matter takes that description substantially closer to the imagery of creation. This is pair production, a process in which energy is converted directly into a pair of particles of opposite charges and equal mass. The process is a familiar one in nuclear laboratories and in cosmic rays; in fact, the first positively charged electron was found as a cosmic ray track. The conventional description of this process is that a particle can be torn out of the vacuum if enough energy is provided, the second particle being the hole left in the vacuum. At terrestrial or stellar densities gravity is not nearly strong enough to mediate pair production, although electromagnetism is. In the first instant of the big bang, when densities were enormous, gravity was intense enough for all the possible elementary particles to be produced by vacuum fluctuations. The present universe is just the cold, transparent, expanded residue of these particles. From this point of view the material universe is thought of as occurring automatically from a sufficiently bumpy vacuum. In the event that the universe produced in this way has enough mass to cause eventual recollapse, the matter would in the last instant return to the vacuum state.

Some theorists have speculated that the physical laws themselves somehow "condense," perhaps statistically, at the same moment that matter appears. This admittedly remote language is reminiscent of Anaximander's dynamic principle, the separation of opposites from the Infinite, which itself echoes the Middle Eastern creation myths. The logical difficulties evident in Anaximander are not absent from the quantum vacuum.

Of the creation myths involving separation, none is more evocative in this context than that of the Egyptians at Heliopolis. The sun-god Atum, standing in the midst of chaos, brings forth from his own body the first pair of gods:

Thou didst spit out Shu
  Thou didst spew out Tefnut.[7]

Shu and Tefnut, air and moisture, from whom Earth and Sky evolve.

There is no reference anywhere to a return of Shu and Tefnut to the body of Atum. This could perhaps be taken as a hint that the missing mass needed to close the universe will not be found. If the Pyramid Texts are right, the universe is open and infinite. On the other hand, many scientists seem to be drawn to a closed and cyclic universe. Is this because we cannot face an infinite past and future, an anxiety not unique to our time (recall the myths of the "Great Year" and the "eternal return")? For the answers, we will put our trust in less personal oracles than the Egyptians did: the hardware of modern astronomy, the space telescope—more distant but less cryptic.

APPENDIX

Experimental Tests of Special and General Relativity

SPECIAL RELATIVITY

1. Time dilation: moving clocks are found to run slower than clocks at rest in a given reference frame.

   a) A muon at rest decays into an electron with a half-life of 2.2 μs (microseconds, 10^{-6} seconds). When moving through the laboratory at 99 percent of the speed of light, the lifetime has been measured to be 15.6 ± 5.8 μs. The observed dilation is by a factor of 7.1; the prediction from $\gamma = 1/\sqrt{(1 - v^2/c^2)}$ is 7.0888.

   b) Pion radiotherapy at Los Alamos Scientific Laboratory requires diverting a pion beam to an experiment or treatment location on the order of 10^2 meters away from where they are produced. The pion lifetime at rest is 2.6×10^{-8} sec. At even 10 percent of the speed of light, the pion would, on the average, travel less than 1 meter—the experiment would be a total failure. At 0.99c the pion would still only travel about 10 m before decaying, were it not for time dilation. On the other hand, at that speed the time dilation factor γ is about 23, and the typical pion can travel more than 200 meters!

   c) Beginning in October 1971, Hafele and Keating flew four cesium atomic clocks around the world in commercial airliners in both the eastward and westward directions. The clocks were expected to experience both special and general relativistic effects totaling -40 ± 23 nanoseconds (a nanosecond is a billionth of a second, i.e., 10^{-9} secs) and $+275 \pm 21$ ns for the eastward and westward trips, respectively. The average observed differences were -59 ± 10 ns, and $+273 \pm 7$ ns, in good agreement with theory. See R. Sexl and H. Sexl, *White Dwarfs-Black Holes*, New York: Academic Press, 1979.

2. Constancy of the speed of light

A variety of tests have been made, from the Michelson-Morley experiment to the use of the Mossbauer effect and a platform rotating at high speed (50,000 revolutions per second). One of the more interesting involved measuring the velocity of light emitted from opposite limbs of the sun. Since the equatorial velocity of the sun is about 2000 m/sec, classically a difference would be expected of 4000 m/sec in the measured velocity; an amount easily measured. In fact the result was consistent with zero: 63 ± 230 m/sec.

3. Relativistic increase in mass.

Early in the development of the particle accelerator called the cyclotron, high enough energies (and hence velocities) were reached so that the relativistic increase in mass created a barrier to further acceleration. For protons, this limit is about 30 MeV (million electron volts of energy). The first measurement was by Buecherer in 1909.

4. Mass-Energy equivalence

Einstein's famous formula $E = mc^2$ is used in every laboratory study of a nuclear reaction and is at the heart of the nuclear reactor and fission and fusion weapons. The theory of stellar energy production given in chapter 15 depends on the validity of this formula. It is an integral part of the special theory of relativity.

GENERAL RELATIVITY

1. Bending of starlight

Rather thoroughly described in chapter 13: the tests have been of the deflection of starlight near the sun, comparison of the angular separation of the quasars 3C273 and 3C279 (the latter being occulted each October 8), the discovery of a "double-quasar 0957 + 561, a manifestation of gravitational focusing, that is, a "gravitational lens" effect.

2. Time delay

Described in chapter 13. In the case of Mercury near superior conjunction (on the other side of the sun from the earth) the time delay is about 240μs, verified to within about 10 percent. In the case of the pulsar CP 0952, which is about 5° from the ecliptic, the effect is 50μs.

3. The perihelion advance

The table that follows, from Sexl and Sexl, shows the results for the solar system. For the pulsar PSR 1923 + 16 the advance is 4° per year.

Planet	Mean Distance From Sun (in millions of km)	Predicted Perihelion Advance	Observed
Mercury	57.91	43.03″	43.11 ± 0.45
Venus	108.21	8.6″	8.4 ± 4.8
Earth	149.60	3.8″	5.0 ± 1.2
Icarus	161.0	10.3″	9.8 ± 0.8

4. Gravitational red-shift

Pound and Rebka (1960) (chap. 13, note 34) verified the general relativistic prediction to within 1 percent. This is, however, the weakest test of general relativity. A related test involves the retardation of clocks in gravitational fields: corrections due to the fact that the National Bureau of Standards in Boulder is at 5400 ft above sea level, while Greenwich is at 80 feet, amounts to more than $1\mu s$ per year—again an easily measured amount.

5. Gravitational radiation

There has been no unequivocal detection of gravitational radiation, but it requires increasingly sensitive measurement. On the other hand, the binary pulsar PSR 1923 + 16 (the "Hulse-Taylor" binary pulsar, period .059 sec) has an orbital period of about 8 hours, which has been shown to be decreasing by about 10^{-5} seconds per year, evidently as the result of the radiation of energy in the form of gravitational waves.

Notes

1. Overview

1. Though William Blake said, "To generalize is to be an idiot. To particularize is the alone distinction of Merit." See W. Frye, *Fearful Symmetry* (Princeton: Princeton University Press, 1947).
2. A. A. Warner, Dean Morse, and T. E. Cooney, eds., *The Environment of Change* (New York: Columbia University Press, 1969), p. 49.
3. Loren Eiseley, *The Firmament of Time* (New York: Atheneum, 1970), p. 176.
4. Adrienne Rich, from the poem "Ghazals" (1968), in *Leaflets, Poems 1965–1968,* (New York: Norton).
5. e. e. cummings, "XXXIII" from *is 5* (1926), in *Complete Poems* (New York: Harcourt Brace Jovanovich, 1972).
6. Thomas Kuhn, *The Structure of Scientific Revolutions,* 2nd enlarged ed. (Chicago: University of Chicago Press, 1970); Karl Popper, *The Logic of Scientific Discovery* (New York: Harper & Row, 1965); S. Toulmin, *The Philosophy of Science: An Introduction* (New York: Harper, 1960; Paul Feyerabend, *Against Method* (London: Humanities Press, 1975), an iconoclastic statement of Feyerabend's "anarchistic" philosophy of science; B. A. Brody and N. Capaldi, eds., *Science: Men, Methods, Goals* (New York: Benjamin, 1968), a very useful collection of readings in the philosophy of science.

2. Earliest Awareness

1. H. Frankfort et al. *Before Philosophy* (Baltimore: Penguin), p. 16. Originally published as *The Intellectual Adventure of Ancient Man* (Chicago: University of Chicago Press, 1946).
2. *Ibid.,* p. 13.
3. *Ibid.,* p. 29.
4. From the poem "Epistle To Be Left in the Earth," by Archibald Macleish.
5. Frankfort et al, p. 15.
6. G. de Santillana and H. von Dechend, *Hamlet's Mill,* p. 166.
7. The so-called "Great Year of the Pleiades." This asterism was very near the vernal equinox a little before 1700 b.c. See Mary Barnard, *The Mythmakers* (Athens: Ohio University Press).
8. In de Santillana and von Dechend, p. 325.

9. H. A. T. Reiche, "The Language of Archiac Astronomy: A Clue to the Atlantis Myth?" In K. Brecher and M. Feirtag, eds. *Astronomy of the Ancients.* (Cambridge: MIT Press, 1979).

10. A Heidel, *The Babylonian Genesis* (Chicago: University of Chicago Press, 1951). J. B. Pritchard, ed., *Ancient Near Eastern Texts Relating to the Old Testament,* E. A. Speiser, trans. (Princeton, N.J.: Princeton University Press, 1958), vol. 1.

11. Frankfort et al., pp. 185–86.

12. *Ibid.,* p. 58. S. N. Kramer, *Mythologies of the Ancient World* (New York Doubleday Anchor, 1961).

3. First Astronomy

1. These discoveries, still somewhat controversial, have been made by Alexander Marshack and are detailed in his *Roots of Civilization* New York: McGraw-Hill, 1972.

2. S. Toulmin and J. Goodfield, *The Fabric of the Heavens* (New York: Harper Torchbook, 1961).

3. A star is said to rise heliacally (with the sun) when it rises just before dawn, that is, just before the rising sun makes it invisible. Every star rises heliacally once a year, except for those that are circumpolar (never set).

4. The earliest Egyptian calendar was in fact lunar, the month beginning with the disappearance of the *waning* crescent. See R. A. Parker. "Ancient Egyptian Astronomy," in F. R. Hodson, ed., *The Place of Astronomy in the Ancient World* (London: Oxford University Press, 1974).

5. O. Neugebauer, *The Exact Sciences in Antiquity* (New York: Dover, 1969). Note however its origin: "In a game of dice, so the story goes, Mercury won five days from the lunar year and added them to the solar year" (resulting in the present values of 355 and 365 days, respectively). H. A. T. Reiche, "The Language of Archaic Astronomy: A Clue to the Atlantis Myth?" in K. Brecher and M. Feritag, eds., *Astronomy of the Ancients* (Cambridge, Mass.: MIT Press, 1979).

6. Since $8^h/24^h = 1/3$ and $360/3 = 120°$ or 12 decans, the night was divided into 12 decans. At the summer solstice and 30° latitude, the interval from the end of evening twilight to the beginning of morning twilight is about 8 hours. But see Parker in "Ancient Egyptian Astronomy," p. 55.

7. See especially Neugebauer, *The Exact Sciences in Antiquity;* O. Neugebauer, *A History of Ancient Mathematical Astronomy,* 3 vols. New York: Springer-Verlag, 1975; and B. L. Van der Waerden, *Science Awakening,* 2 vols. (Leyden: Noordhoff, 1974–75).

8. R. C. Thompson, *The Reports of the Magicians and Astrologers of Ninevah and Babylon,* p. 208, quoted in A. Pannekoek, *A History of Astronomy* (New York: Wiley Interscience, 1961), p. 39.

9. O. Neugebauer, *The Exact Sciences in Antiquity,* p. 71. In a similar vein is Parker's comment, "Throughout the three millennia of recorded Egyptian history, we have nothing whatever to suggest that the movements of the Moon and planets were systematically observed and recorded as they were in Babylonia." Parker, "Ancient Egyptian Astronomy."

10. Some would argue that it pales by comparison with the riches of nearby Avebury, whose astronomical significance, is however, unknown.

11. Recalibration of the C^{14} dating technique has pushed the dates of Stonehenge back more than 600 years compared to those assumed in the early 1960s.

12. The periodic variation in inclination of the earth's spin axis is called nutation.

13. Typical of the initial reaction are papers by R. J. C. Atkinson, "Moonshine on Stonehenge," *Antiquity* (1966), 40:212; and by Jacquetta Hawkes, "God in the Machine," *Antiquity* (1967), 41:174.

14. C. A. Newham, *The Astronomical Significance of Stonehenge* (Leeds: John Blackburn, 1972).

15. G. Hawkins, "Stonehenge, a Neolithic Computer," *Nature* (1964), 202:1258.

16. Fred Hoyle, *On Stonehenge* (San Francisco: Freeman, 1977).

17. A. Thom, *Megalithic Lunar Observatories.* (Oxford: Clarenden, 1971)

18. The Pleiades or "seven sisters" are a bright cluster rising just ahead of Orion and Sirius in the winter sky.

19. J. Eric S. Thompson, "A Commentary on the Dresden Codex," *Memoirs of the American Philosophical Soc.* (1972), 93, Philadelphia.

4. The Fountainhead

1. The poetry of Hesiod, from the eighth century B.C., especially his *Theogony,* provides to some degree a transition between the personal, animistic cosmologies of Babylon and Egypt and the rational cosmologies of the pre-Socratics.

2. G. S. Kirk and J. E. Raven, *The Presocratic Philosophers* (Cambridge: Cambridge University Press, 1957). A good source for the pre-Socratics; translations of original sources and commentaries.

3. Oxygen should be as pungent as sulfur; they are similar chemical agents. It is a triumph of natural selection that the world does not have a bad odor.

4. The moon moves around the earth in just this way, with the same side always facing the center of its orbit. This "rolling" motion became a common feature in geocentric planetary models, as we shall see.

5. Aristotle, *On Democritus.* See Kirk and Raven, p. 418.

6. *Ibid,* p. 173.

7. de Santillana, *Origins of Scientific Thought* (Chicago; University of Chicago Press, 1961), p. 35.

5. Synthesis

1. F. M. Cornford, *Plato's Cosmology: The Timaeus of Plato* (New York: Humanities Press, 1952). See also the conclusion to Plato's *Republic.*

2. Thomas S. Kuhn, *The Copernican Revolution* (New York: Vintage, 1957), chapter 3.

3. Aristotle, *On the Heavens* (De Caelo), W. K. C. Guthrie, trans. (London: Wm. Heinnemann, Ltd., 1934), p. 23.

4. Despite the (correct) inference that the slower planets are more distant from the earth, the Greeks were characteristically undogmatic about the arrangement of the planets, many placing the sun midway (fourth among seven). See O. Neugebauer, *A History of Ancient Mathematical Astronomy,* 3 vols. (New York: Springer-Verlag, 1975), *Part II,* p. 690; and E. Grant, *A Source Book in*

Medieval Science (Cambridge: Harvard, University Press), 1974. Uranus, incidentally, was discovered in 1781, Neptune in 1845–46, and Pluto in 1930.

5. Eudoxus' system was fully described by S. Schiaparelli in 1874. It was Schiaparelli who first characterized mathematically the "hippopede." See, for example, J. L. E. Dreyer, *A History of Astronomy from Thales to Kepler* (New York: Dover, 1953), chapter 4. Simplicius' work was a commentary on Aristotle's *De Caelo*.

6. Dreyer, p. 93. Neugebauer says that the surprising assumption that the sun wanders from the ecliptic may be traceable to the "theoretical" ecliptic of Pythagorean origin, inclined at 24° to the celestial equator, or *exactly* 1/15 part of a great circle. Neugebauer, p. 629.

7. Dreyer, p. 107.

8. *Ibid,* p. 101.

6. Orb in Orb

1. The truly astounding "Antikythera Mechanism," found off the coast of Crete in 1900–1 ought to be mentioned. Because it is unique, its importance in our discussion of Greek science is unclear. Certainly it reveals a level of Greek technology heretofore unsuspected. Dating from before 50 B.C., it is a complex clockwork functioning as a calendrical sun and moon computer. See D. J. de Solla Price, *Science Since Babylon* (New Haven, Conn.: Yale University Press, 1961).

2. T. L. Heath, *The Works of Archimedes*. Dover reprint of 1897 edition, with 1912 supplement containing Archimedes *Method,* one of the remarkable documents of the history of science.

3. *Ibid.,* p. 221.

4. A Pannekoek, *A History of Astronomy* (New York: Wiley Interscience, 1961), p. 121.

5. Aristarchus, "On the Sizes and the Distances of the Sun and Moon," T. L. Heath, trans., reprinted in M. Cohen and I. E. Drabkin, *A Source Book in Greek Science* (New York: McGraw-Hill, 1948), p. 109. See also Pannekoek, p. 119.

6. Which may, in fact, have been a comet.

7. The "line of apsides" connects the nearest point in the moon's orbit around the earth to the farthest point.

8. Planetary longitudes are analogous to longitudes on the earth; they specify planetary positions parallel to the ecliptic. Planetary latitudes measure positions away from the ecliptic due to the inclinaton or tilt of the orbits of the planets compared to the plane of the earth's orbit.

9. Ptolemy does, in his later work *Hypothesis,* seem to commit himself, after all, to the reality of his devices.

10. R. Newton, *The Crime of Claudius Ptolemy* (Baltimore: Johns Hopkins University Press, 1977). See, however, the review by J. D. North in *American Scientist* (July–August, 1978), p. 511, and the very long analysis by Owen Gingerich, *Quarterly Journal of the Royal Astronomical Society* (1980), 21:253.

11. J. L. E. Dreyer, *A History of Astronomy from Thales to Kepler* (New York: Dover, 1953), p. 107.

12. Theon of Smyrna, *Astronomia*. Quoted in P. Duhem, *To Save the Appearances* (Chicago: University of Chicago Press, 1969).

13. Ptolemy, *The Almagest,* 13.2 Quoted in Duhem, *ibid.,* p. 17.

14. Proclus, *Hypotyposes*. In Duhem, *ibid.*, p. 19.

15. *Ibid.*, p. 21.

16. Thomas Aquinas, *Summa Theologica* (New York: Benziger Bros., 1947–48).

7. Medieval Europe

1. Pliny, *Natural History*. There are several translations available, including one by H. Rackham in the Loeb Classical Library. Pliny died in A.D. 79, during the eruption of Vesuvius.

2. E. Brehaut, *An Encyclopedist of the Dark Ages* (New York: Columbia University Press, 1912). Quoted in E. Grant, *A Source Book in Medieval Science* (Cambridge: Harvard University Press, 1974), p. 3.

3. J. L. E. Dreyer, *A History of Astronomy from Thales to Kepler* (New York: Dover, 1953), p. 211.

4. Lactantius, *Divine Institutions*. See Dreyer, p. 209.

5. See M. Clagett, *Greek Science in Antiquity* (New York: Abelard-Schuman, 1955), ch. 13. Earlier Plutarch had expressed some quite remarkable and very non-Aristotelian ideas about gravity in "On the Face of the Moon," apparently derived from the Stoic Poseidonius. See S. Sambursky, *The Physical World of the Greeks*, M. Dagut, trans. (London: Routledge and Kegan Paul, 1956), p. 206.

6. Algebra, for example, and the names of the bright stars Aldebaran, Betelgeuse, and Rigel.

7. E. Grant, *A Source Book in Medieval Science*, p. 3. Originally published in L. Thorndike, *University Records and Life in the Middle Ages* (New York: Columbia University Press, 1944).

8. For example, Aquinas' personal translator, William Moerbeke, translated nearly all of Archimedes' works into Latin, making them finally available to the Middle Ages. Their influence was enormous.

9. A. C. Crombie, *Augustine to Galileo: The History of Science, A.D. 400–1600* (Cambridge: Harvard University Press, 1961), 2:1.

10. The influence of medieval technology on the rise of experimental science is often neglected. Specifically the invention of the reciprocating crank and connecting rod, the treadle, the spinning wheel, spectacles, and medieval and gothic architecture. See L. Thorndike, *A History of Magic and Experimental Science*, 8 vols. (New York: Macmillan and Columbia University Press, 1923–1958).

11. Crombie, *Augustine to Galileo*, vol. 2:23.

12. Koyre has disputed the revolutionary character of this development and questioned the importance of methodology in scientific discovery. See A. Koyre, "The Origins of Modern Science: A New Interpretation," *Diogenes* (1956), 16:1.

13. Occam died in the plague that ravaged Europe in 1348–49. Jean Buridan probably succumbed in the next outbreak nine years later.

14. In modern terms, momentum is mass time velocity (mv, a *vector* quantity), while kinetic energy is one-half the mass times the square of the velocity ($\frac{1}{2}mv^2$, a *scalar*).

15. E. Grant, *A Source Book in Medieval Science*, pp. 276–77, reprinted from M. Clagett, *The Science of Mechanics in the Middle Ages* (Madison: University of Wisconsin Press, 1959).

16. A. D. Menut and A. J. Denomy, *Oresme's Le Livre du ciel et du Monde*

(Madison, Wisc.: University of Wisconsin Press, 1968). See also reference 17.

17. See selections in E. Grant, *A Source Book in Medieval Science.* The efforts of Albert of Saxony (c.1357) to understand accelerated motion are especially notable. He debated whether the speed of fall was proportional to the distance fallen or to the elapsed time (correctly) but ended up rejecting both ideas. Perhaps the first to argue persuasively that falling bodies accelerate was Strato, known as "the physicist," who flourished about 280 B.C. He is said to have argued from the way a descending stream of water disintegrates as it falls.

18. The role of Leonardo Da Vinci is especially interesting. He epitomizes the problems facing the man of genius in this period of transition from medieval to modern mechanics. We see him somewhat unsystematically wrestling with fundamental problems of mechanics and being unable to bring any order to it. See E. J. Dijksterhus, *The Mechanization of the World Picture* (Oxford: Clarendon, 1961)—one of the best studies of the foundations of modern science.

19. R. T. Gunther, *Early Science in Oxford* (Oxford: Oxford University Press, 1928), vol. 2, p. 288.

20. The point at which the sun is most distant from the earth.

21. It had been translated anonymously directly from Greek into Latin in Sicily in 1160.

22. Even the Koran was translated into Latin, in 1142!

23. Whose *Epitome of the Almagest* may have strongly influenced Copernicus.

24. The late fourteenth century saw a decline of the University of Paris, related to the ravages of the plagues, the papal schism of 1378, and the hundred-years war. The faculty at Paris fled to Italian universities. At the same time, Wycliff rejected ecclesiastical authority and the Bible was translated into English from the Latin for the first time. Europe was coming into the "modern world."

8. The Copernican Revolution, 1

1. Nicolaus Copernicus, *On the Revolutions of the Heavenly Spheres,* Book I. See Thomas S. Kuhn, *The Copernican Revolution* (New York: Vintage, 1957), p. 154.

2. There were, of course, other important figures, Bacon, Huygens, and Descartes, for example, and we shall hear more of them. But in the history of cosmological thought the crucial roles of the figures we have mentioned is clear.

3. Thomas Kuhn, *The Structure of Scientific Revolutions,* 2nd enlarged edition (Chicago: University of Chicago Press, 1970).

4. A. D. White, *A History of Warfare of Science with Theology in Christendom* (New York: Appleton, 1896), vol. 1, p. 126.

5. *Commentariolus,* in *Three Copernican Treatises,* Edward Rosen, trans. (New York: Columbia University Press, 1939), p. 57.

6. The question of the influence of the Maragha school on Copernicus is an interesting and important one but cannot be pursued here. See ref. 19.

7. Nicholaus Copernicus, *On the Revolutions of the Heavenly Spheres,* C. G. Wallis, trans. *Great Books of the Western World,* 16:507.

8. From "Discoveries" by Vernon Watkins. It is true that Ptolemy's lunar theory gave a very poor value for the ratio of the apogee distance to perigee distance, but the longitudes were given with astonishing accuracy.

9. In *de Revolutionibus* of 1543 there is no mention of Aristarchus. But in the manuscript, both he and Pythagoras are mentioned.

10. *On the Revolutions,* p. 521.

11. *Commentariolus,* p. 63.

12. *On the Revolutions,* p. 526.

13. *Ibid,* p. 516.

14. S. Toulmin and J. Goodfield, *The Fabric of the Heavens* (New York: Harper & Row, 1961), p. 124.

15. Otto Neugebauer has remarked that not a page of *de Revolutionibus* can be understood without a knowledge of *The Almagest.*

16. Copernicus, *On the Revolution of the Heavenly Spheres,* in *Great Books of the Western World,* 16:664.

17. *On the Revolutions,* p. 505.

18. *Ibid,* p. 509.

19. Libration is motion back and forth along a straight line, so that the path "describes certain lines similar to a twisted garland." That rectilinear motion can be compounded from two equal, counterrotating circles had been demonstrated by the Persian astronomer Nasir-al-Din of Tus, whose influence on Copernicus is, however, unknown. See W. Hartner in *Proceedings of the American Philosophical Society* (1973) 117:413.

9. The Copernican Revolution, 2

1. Copernicus, *On the Revolution of the Heavenly Spheres,* in *Great Books of the Western World,* 16:664.

2. The Tychonic system is mathematically equivalent to the Copernican system. When Galileo discovered that the phases of Venus are incompatible with the Ptolemaic system, the Tychonic system was the only alternative available to those unable to embrace the heliocentric hypothesis.

3. A. Koestler, *The Watershed* (Garden City, N.Y.: Doubleday, 1960), is the massive chapter on Kepler from *The Sleepwalkers,* published separately (New York: Grosset and Dunlap, 1959). See *Watershed,* p. 109 or *Sleepwalkers,* p. 302.

4. When Kepler arrived in Prague, he bet Longomontanus that he could solve the theory of Mars within a week. (O. Gingerich, "Johannes Kepler and the New Astronomy," *Quarterly Journal of the Royal Astronomical Society* (1972), 13:346. It took him five years.

5. See the Bibliography, especially Caspar's *Kepler* and Beer and Beer. Note that Kepler's *Astronomia Nova* was published in 1609, the very year Galileo began his telescopic discoveries. Without question, the "new astronomy" dates from 1609.

6. For example, the passage quoted by T. Kuhn in *The Copernican Revolution* (New York: Vintage, 1957), p. 131, derived from E. A. Burtt, *The Metaphysical Foundations of Modern Science,* 2d ed. (New York: Harcourt Brace, 1932).

7. More precisely, it is a statement of conservation of angular momentum.

10. Galileo

1. It is known that Simon Steven dropped weights from a height of 30 feet in 1586 in such an experiment. Our information that Galileo actually dropped balls from the Tower of Pisa, which he may very well have done in some sort of public demonstration, comes from an account by Vincenzio Vivian, written

nearly 50 years after the event. See S. Drake, *Galileo at Work* (Chicago: University of Chicago Press, 1978).

2. Drake, in *Galileo at Work,* argues that Galileo abandoned his belief in the Copernican theory between the late 1590s and 1609–10, because of the failure to observe parallax. According to this argument he embraced it again following his discoveries with the telescope. For a translation of the "Letters on Sunspots," see S. Drake, *Discoveries and Opinions of Galileo* (Garden City, N.Y.: Doubleday Anchor, 1957), pp. 86–1144.

3. Toricelli worked, however, with Galileo only during the three months preceding the latter's death. He did carry on Galileo's work, and he succeeded him as mathematician to the Grand Duke of Tuscany. He had been a student of Galileo's great student, Benedetto Castelli. Castelli died the year after Galileo; Cavalieri, the great mathematician and correspondent of Galileo, student of Castelli, died in 1647, as did Torricelli.

4. Especially Christopher Clavius.

5. L. Geymonat, *Galileo Galilei,* S. Drake, trans. (New York: McGraw-Hill, 1965), p. 75.

6. *Ibid,* p. 85.

7. *Ibid,* p. 92.

8. The two appendices to Drake's translation of Geymonat's *Galileo Galilei* provide contrasting views on the matter, one by Drake and the other by G. de Santillana. See also de Santillana's *Crime of Galieo* (Chicago: University of Chicago Press, 1955) and Drake's *Galileo at Work.*

9. Bellarmine died in 1621.

10. Drake, *Galileo at Work,* p. 349.

11. Geymonat, *Galileo Galilei,* p. 153.

12. de Santillana, *The Crime of Galileo,* p. 310.

13. Geymonat, *Galileo Galilei,* p. 153–154.

14. Though Francesco Barberini, who served Galileo's interest throughout the proceeding, had forbidden torture, a fact unknown, of course, to Galileo.

15. de Santillana, *The Crime of Galileo,* p. 324.

16. Galileo's arguments for the motion of the earth, based on the paths taken by sunspots across the sun and on the tides, were both mistaken. Bradley, 1729, while riding in the rain on a boat in the Thames, was struck by the explanation of the 40″ per year annual motion of the stars (stellar aberration), which had been first observed in 1640. It was not until 1837 that Bessel measured the 0.3″ parallax of the star 61 Cygni.

17. Drake, *Discoveries and Opinions of Galileo,* p. 182.

18. A. Koyre, *Galileo Studies,* John Mepham, trans., Atlantic Highlands, N.J.: Humanities Press, 1978. Quoted in Geymonat, *Galileo Galilei,* p. 132–133.

19. *Dialogue on the Two Great World Systems,* Salusbury translation, revised by G. de Santillana (Chicago: University of Chicago Press, 1953), p. 199.

20. *Ibid.,* p. 128.

21. *Dialogue Concerning the Two Chief World Systems,* Stillman Drake, trans. (Berkeley: University of California, 1967), p. 464.

22. Including, apparently, a naked-eye observation of Kepler, who thought he was seeing a transit of Mercury. Johann Fabricius had made the earliest telescopic observations of sunspots in 1610, several months before Galileo or Scheiner. See Drake, *Galileo at Work,* p. 213.

23. The Tychonic system will yield the correct phases for Venus, but it, even in the seventeenth century, was an improbable construction.

24. H. Dingle, *The Scientific Adventure* (New York: Philosophical Library, 1953), p. 51.

25. de Santillana, *The Crime of Galileo*, p. 9.

26. Letter to Fra Paolo Sarpi, Oct. 16, 1604, quoted in Geymonat, *Galileo Galilei*, p. 29.

27. On the other hand, speed in this quotation is a rendering of the imprecise *velocita*, which was *ipso facto*, proportional to distance. Thus it is v^2, proportional to the kinetic energy. Galileo had not yet arrived at an understanding of our idea of instantaneous speed. See Drake, *Galileo at Work*, p. 102.

28. S. Drake, *Galileo Studies* (Ann Arbor, Mich.: University of Michigan Press, 1970).

29. *Ibid.*

30. Drake, *Discoveries and Opinions of Galileo*, p. 237.

31. Along with his correspondent, Bonaventura Cavalieri.

32. S. Drake, trans., *Dialogue Concerning the Two Chief World Systems*, p. 216.

11. The Newtonian Synthesis

1. See, for example, C. C. Gillispie, *The Edge of Objectivity* (Princeton, N.J.: Princeton University Press, 1960), p. 117.

2. Of course, as Newton said (the statement was not original): ". . . I have stood on the shoulders of giants." Some of these giants were Kepler, Galileo, Descartes, Wren, Huygens, and Hooke. The latter, a man of genius himself, is in the end a tragic figure. Utterly defeated by Newton, he lies in an unmarked grave.

3. A. Koestler, *The Sleepwalkers* (New York: Grosset and Dunlap, 1963).

4. Only because in 1642, England was still on the Julian calendar, while Catholic Europe was using the new (1582) Gregorian calendar.

5. See, for example, I. B. Cohen, "Newton's Discovery of Gravity," *Scientific American* (March 1981), 244:166.

6. It is possible, perhaps likely, that Hooke, who might have been regarded as the giant of the seventeenth century but for Newton, may have provided the latter with the crucial insight, concerning centripetal acceleration toward the center of an orbit as against inertial motion in the tangential direction. Cohen, "Newton's Discovery." See also R. Westfall, "Hooke and the Law of Universal Gravitation," *British Journal for the History of Science* (1967), 3:245, 251; and L. Rosenberg, "Newton and the Law of Gravitation," *Archive for the History of the Exact Sciences* (1965), 2:375. In his "Newton in the Light of Recent Scholarship," *Isis* (1960), 51:489, Cohen quotes Newton's biographer Stukeley: "After dinner, the weather being warm, . . . he and Newton went into the garden and drank tea, under the shade of some appletrees. . . ." There Newton told Stukeley that "he was just in the same situation; as when formerly, the notion of gravitation came into his mind. It was occasioned by the fall of an apple, as he sat in a contemplative mood."

7. See B. J. T. Dobbs, *The Foundations of Newton's Alchemy* (Cambridge: Cambridge University Press, 1975). On Newton's theological perceptions, see the biographies by More, Manuel, and Westfall.

8. Almost as a swan song, he solved a famous problem posed by Leibniz and Johann Bernoulli, and thereby invented a new field of mathematics, the calculus of variations. See, for example, Hoyle's version of the story, an enchanting one, in his *Astronomy and Cosmology* (San Francisco: Freeman, 1975), p 431. This was in early 1697.

9. A. R. Hall. *Philosophers at War: The Quarrel Between Newton and Leibniz* (New York: Cambridge University Press, 1980).

10. Acceleration is defined as the time rate of change of velocity. Symbolically, $a = v_2 - v_1/t_2 - t_1 = \Delta v/\Delta t$, where $\Delta v =$ change in velocity and $\Delta t; =$ time interval during which the change occurs.

11. Cohen, "Newton's Discovery," and his "Newton in the Light of Recent Scholarship," refs. 5 and 6. Cohen's current view seems to be that Newton concocted the whole story of his discovery of gravitation in 1665–66 in order to make it appear that the origins of universal gravitation came 20 years earlier than they did. On the other hand, see the definitive Newton biography, *Never at Rest*, by Westfall on this particular question.

12. Newton, *Mathematical Principles of Natural Philosophy*, A. Motte, trans.; revised by Florian Cajori (Berkeley: University of California Press, 1934).

13. *Ibid.*, p. 419.

14. *Ibid.*, p. 397.

15. *Ibid.*, p. 406.

16. *Ibid.*, p. 546.

17. *Ibid.*, p. 547.

18. Newton did participate in the invention of the reflecting telescope, and he was interested in the application of astronomy to problems of navigation at sea.

19. See A. Koyre, *Newtonian Studies* (Chicago: University of Chicago Press, 1965), p. 63.

12. Widening Horizons

1. A good example is the comment by Minkowski paraphrased by Guth, that he was surprised when he learned of Einstein's seminal work in special relativity, because "he was not such a good student in Zurich." E. Guth, "Einstein's geometry as a branch of physics," in *Relativity*, M. Carmeli, S. I. Fickler, and L. Witten, eds (New York: Plenum, 1970), p. 178.

2. The term "field" is used to describe a region of physical influence, specified quantitatively by a mathematical function also called the "field." Changes in the field functions are specified by "field equations."

3. The term itself ("field") was not in use in Newton's time.

4. Quoted in *The Atomists*, by B. F. J. Schonland (Oxford: Oxford University Press, 1968).

5. "Autobiographical Notes," in P. A. Schilpp, ed. *Albert Einstein: Philosopher-Scientist* (New York: Harper and Row, 1959), p. 17.

6. The *annus mirabilis*, 1905, when Einstein invented special relativity and the quantum of light and explained Brownian motion, has captured the imagination of biographers and historians. Martin J. Klein has discussed the connections among these apparently disparate discoveries. See, for example, "Einstein and the Development of Quantum Physics," in A. P. French, ed., *Einstein, A Centenary Volume* (Cambridge, Mass.: Harvard University Press, 1979), pp. 133–

151. The Nobel Prize was awarded in 1921, ostensibly for the work on the photoelectric effect that led to the light quantum, since relativity was still controversial.

7. A. Einstein, "On the Electrodynamics of Moving Bodies," *Annalen der Physik,* Series 4, (1905), 17:891.

8. For example, A. Einstein, *Relativity,* R. W. Lawson, trans. (New York: Crown, 1961).

9. *Ibid.,* pp. 25–27.

10. *Ibid,* p. 36.

11. The experimental results obtained by Walter Kaufmann in 1906 for electrons at relativistic speeds seemed to refute Einstein's predictions. Einstein was characteristically unperturbed, and later measurements (those of A. H. Bucherer in 1909 were the first) demonstrated the correctness of Einstein's results.

12. In recent years distant radio sources have been observed with features that are expanding at velocities of up to 50 c (50 times the velocity of light), far in excess of the maximum possible velocity. These "supraluminal velocities" can be explained if the particles making up the source are moving in a relativistic jet nearly along the line of sight.

13. At an acceleration of 1 g, for example, which would do no more than give the passengers a feeling of their natural weight (i.e., on earth), a trip to the Andromeda Galaxy, M31, which would take more than 2 million years earth time, would take only 28 years from the point of view of the passengers. The energy cost would, of course, be enormous. See, for example, I. S. Shklovskii and Carl Sagan, *Intelligent Life in the Universe* (New York: Holden-Day, 1966), chapter 32.

13. General Theory of Relativity

1. H. Minkowski, "Space and Time," translated in H. A. Lorentz, A. Einstein, H. Minkowski, and H. Weyl, *The Principle of Relativity,* W. Perrett and G. B. Jeffery, trans. (New York: Dover, 1923).

2. For a popular account, see T. Ferris, *The Red Limit* (New York: William Morrow, 1977).

3. Edwin P. Hubble, "A Relation Between Distance and Radial Velocity Among Extragalactic Nebulae," *Proceedings of the National Academy of Sciences, U.S.* (1929), 15:169.

4. A fact quite evident from a reading of the early (1907–16) papers, e.g., those translated in Lorentz et al, *The Principle of Relativity.* Included is the groundbreaking 1917 paper "Cosmological Considerations in the General Theory of Relativity," in which Einstein's concern about whether the universe is finite or infinite is discussed strictly as a mathematical boundary value problem. In the end, the universe is assumed to be finite because of the difficulty in formulating proper boundary conditions at infinity. He concludes, "At any rate this view is logically consistent, and from the point of view of the General Theory of Relativity lies nearest at hand; whether, from the standpoint of present astronomical knowledge, it is tenable, will not here be discussed."

5. See chapter 12. Some, in fact, consider Henri Poincaré to have been the founder of special relativity.

6. Max Born, quoted in R. W. Clark, *Einstein, The Life and Times* (New York: World, 1971), p. 200.

7. See H. A. Lorentz et al, *The Principle of Relativity*.

8. The rotating earth is not an inertial frame. If we apply Newton's second law to motion on the earth's surface, we find that the force (in $F=ma$) has to be modified to include the centrifugal and coriolis forces.

9. This statement is called by Dicke the "weak principle of equivalence" (see note 20).

10. To be precise, the *principle of equivalence* is true only "locally"; that is, a gravitational field can be simulated or replaced by an accelerated coordinate system only at a point, or in practice, in a small region. If two balls are dropped in an elevator fixed in the earth's gravitational field, they will slowly approach each other since the acceleration of each is directed toward the center of the earth. In an accelerated frame (elevator) this does not happen. Thus a single acceleration cannot globally replace the effect of the earth's gravitational field, whose direction varies from point to point over its surface.

11. If the idea of "force" is retained, then because of the tensor character of the field, force (rate of change of momentum) and acceleration will have different directions. See J. D. North, *The Measure of the Universe* (Oxford: Clarendon, 1965).

12. A. Einstein, *Journal of Radioactivity* (1907), 4:440.

13. Translated in H. A. Lorentz et al, *The Principle of Relativity,* p. 117.

14. "On the Influence of Gravitation on the Propagation of Light," *Ibid.,* pp. 97–108. This calculation was not fundamentally different from that of Soldner in 1801. See E. Guth, "Contribution to the History of Einstein's Geometry as a Branch of Physics," in M. Cammeli, S. I. Fickler, and L. Witten, *Relativity* (New York: Plenum, 1970).

15. A. Einstein and M. Grossmann, *Zeit. fur Math. Phys.* (1914), 63:215.

16. A. Einstein, *Preuss. Akad. Wiss. Berlin, Sitzber,* (Nov. 11, 1915), pp. 778–786; (Nov. 18, 1915), pp. 799–801; and (Nov. 25, 1915) pp. 831–839.

17. A. Einstein, *Autobiographical Notes,* P. A. Schilpp, trans. (La Salle, Ill.: Open Court, 1979). First published in 1949 in *Albert Einstein: Philosopher-Scientist,* P. A. Schilpp, ed.

18. Ernst Mach expressed this idea, that inertia is due to interaction with the stars, in his monumental *The Science of Mechanics* (London: Open Court, 1942). Bishop George Berkeley, in *De Motu* (1721), was apparently the first to talk about such a connection between inertia and the large-scale properties of the universe. Brans and Dicke (see chapter 14) explicitly incorporated Mach's principle into their scalar-tensor theory of gravitation, whereas V. I. Lenin felt called upon to reject it on philosophical grounds. Einstein coined the term "Mach's Principle" but rejected Mach's positivism. For his part, Mach came to reject the principle named after him.

19. The Foucault pendulum consists of a very heavy mass swinging freely on a very long wire. Its oscillations define a plane in space, which, as the earth rotates, appears to turn relative to the building in which it is situated. At 30° latitude the plane of the pendulum rotates once in 48 hours; at the poles the period is 24 hours. Foucault pendula can be found in many natural history museums and planetariums.

20. R. H. Dicke, "The Many Faces of Mach," in Hong-Yee Chiu and W. F. Hoffmann, eds., *Gravitation and Relativity* (New York: Benjamin, 1964).

21. An interesting and accessible introduction to non-Euclidean geometry is R. Y. B. Rucker, *Geometry, Relativity, and the Fourth Dimension* (New York:

Dover, 1977). This is a delightful book that owes a lot to E. A. Abbott's *Flatland*. It has an excellent annotated bibliography that will tempt the reader, not only with the likes of Einstein and Wheeler, but Borges and Castenada as well (to mention only two).

22. Surveyors, of course, use this principle without giving it a thought. Now that surveying is making increasing use of portable lasers, the assumption must be increasingly obvious.

23. In 1821–23 Gauss surveyed a triangle with sides of 69, 85, and 107 km defined by three mountain tops, in an unsuccessful attempt to detect a departure from Euclidean geometry.

24. N. Lobachevski, quoted by North, *The Measure of the Universe*, p. 76. Lobachevski's attempts to survey a triangle whose base was the diameter of the earth's orbit and whose apex was Sirius led him to conclude that non-Euclidean geometry had no application to nature. See M. Jammer, *Concepts of Space* (Cambridge: Harvard University Press, 1954), p. 148.

25. The pseudosphere, a surface with constant negative curvature, is hornlike in shape.

26. Because of a technical distinction between local and global or topological connectivity, a cylinder is not *equivalent* to a plane.

27. In an elliptical universe, a limit on the radius of curvature could be obtained by noting that the other side of the sun cannot be seen by looking in the diametrically opposite direction! Using Shapley's value for interstellar absorption of 0.01 magnitude per kiloparsec (recall that a parsec is 3.26 light years), Willem de Sitter arrived at a minimum value of 10^6 parsecs for the radius of curvature, valid only if the sun is old enough for light to travel around the universe from it. See North, *The Measure of the Universe*, p. 80.

28. E. F. Taylor and J. A. Wheeler, *Space-time Physics* (San Francisco: Freeman, 1966).

29. From a lecture to the Cambridge Philosophical Society, February 21, 1870, titled "On the Space Theory of Matter." Published in the *Proceedings of the Cambridge Philosophical Society* (1876), 2:157. Also in W. Clifford, *The Common Sense of the Exact Sciences*, J. R. Newman, ed. (New York: Knopf, 1946).

30. See appendix for additional details. Also section III, "Experimental Verification of the General Theory of Relativity," in G. Tauber, *Albert Einstein's Theory of General Relativity* (New York: Crown, 1979). Another source is C. M. Will, "Einstein on the Firing Line," *Physics Today* (October 1972), 25:23.

31. To Eddington, the greatest event of his life.

32. "Multiple Quasars and Gravitational Lenses," *Sky and Telescope* (December 1980, 60:486. Also F. H. Chaffee, "The Discovery of a Gravitational Lens," *Scientific American* (Nov. 1980). (1980) *243*:70.

33. I. I. Shapiro, "Fourth Test of General Relativity," *Physical Review Letters* (1964), 13:789. See also Tauber, *Albert Einstein's Theory of General Relativity*, p. 139.

34. R. V. Pound and G. A. Rebka, "Apparent Weight of Photons," *Physical Review Letters* (1960), 4:337. Also Tauber, p. 132.

35. S. W. Hawking and G. F. R. Ellis, *The Large Scale Structure of Spacetime* (Cambridge: Cambridge University Press, 1973), p. 365. They give a translation of the arguments of Laplace (Pierre Simon, Marquis de Laplace) from 1798, in which he shows that the escape velocity from the surface of a star with radius $2GM/c^2$ would equal the velocity of light, c, using only Newtonian gravity.

Simon Schaffer in "John Mitchell and Black Holes," *Journal for the History of Astronomy* (1979), 10:42, describes the Yorkshire clergyman's demonstration of the same result. Whether Laplace knew of Mitchell's ideas is not known. Mitchell's name also appears in the history of astronomy in his attempts to correct William Herschel's errors in estimating distances in the galaxy.

36. It is possible, even probable, that the fluctuations will make the mathematical singularities of general relativity meaningless. T. Padmanabhan and J. V. Narlikar, *Nature* (1982), 295:677.

37. J. A. Wheeler, in M. Rees, R. Ruffini, and J. A. Wheeler, *Black Holes, Gravitational Waves, and Cosmology* (New York: Gordon and Breach, 1974), p. 290.

14. Primeval Atom

1. This discovery, which was based on Hubble's distance measurements using cepheid variable stars and Slipher's measurements of red shifts, was anticipated by Carl Wirtz in 1922, who used average size to determine distance. Lemaitre, when in 1927 he published his expanding cosmologic models, noted that these solutions were consistent with the radial velocity of extragalactic nebulae; Robertson, one year later, made the same comparison.

2. As mentioned in the previous chapter, the Curtis-Shapley debate over the nature and distance of the "spiral nebulae" was held in 1926, which is the year in which Hubble published a landmark paper on the nearby galaxy (a member of our local cluster) M33 in Triangulum, which all but conclusively established the great distance of these objects. E. P. Hubble, "A Spiral Nebula as a Stellar System: Messier 33," *Astrophysical Journal* (1926), 63:236.

3. "Cosmological Considerations on the General Theory of Relativity," published in the Proceedings of the Prussian Academy of Science, translated in Lorentz et al. *The Principle of Relativity* (New York: Dover, 1923).

4. This model was interesting from another point of view. When particles were added, they receded from each other—a nonstatic universe.

5. "Homogeneous" means uniform, or the same throughout; "isotropic" means the same in all directions.

6. The Doppler effect, named after Christian Doppler (1803–1853), represents the observation that, with sound or light, the wavelength of a signal from a source approaching the observer is shifted to higher frequency (higher pitch in the case of sound, or toward the blue end of the visible spectrum, in the case of light), while a signal from a receding source is lowered in frequency. The physics of the Doppler effect is different for sound and light, and in the latter case it requires special relativity to give a correct explanation.

7. Which is not to say that there is no controversy over interpretation of the red shifts. Among the important astronomers and cosmologists who question the standard interpretation are Halton Arp, Geoffrey Burbidge, Fred Hoyle, and J. Narlikar. A rather technical survey is G. B. Field, H. Arp and J. H. Bahcall, *The Redshift Controversy* (Reading, Mass.: Benjamin, 1973). There is, nonetheless, a strong consensus favoring the universal recession of the distant galaxies.

8. Nicolas of Cusa. In *Of Learned Ignorance* (1440), G. Heron, trans. (London: Routledge and Paul, 1954) Cusa wrote "The fabric of the world has its center

everywhere and its circumference nowhere." The term "cosmological principle" is due to E. A Milne.

9. This brings to mind the passage from Kipling's "The Cat That Walked by Himself": "I am the cat who walks by himself, and all places are alike to me." C. B. Collins and Stephen Hawking have argued that if the universe were not isotropic and homogeneous, life might not exist. See *Astrophysical Journal* (1973), 143:317.

10. A parsec ("parallax of one second") is the distance at which a star, when viewed from the baseline of the earth's orbit, would have a parallax of one second of arc; that is, its position would vary one second either way from its average position during the year, owing to the earth's motion. Alternatively, it is the distance at which the radius of the earth's orbit would subtend an angle of one second of arc. The parsec is 206,265 astronomical units (earth-sun distances) or about 3.26 light-years.

11. See W. Rindler, *Essential Gravitation* (New York: Springer-Verlag, 1977), p. 213; and E. R. Harrison, *Cosmology* (Cambridge: Cambridge University Press, 1981), chap. 10.

12. $z = \Delta\lambda/\lambda$, the fractional change in wavelength.

13. Quasars (quasi-stellar radio sources) are star-like objects showing enormous red shifts, which seem to imply that they are at great distances and are therefore the youngest or most primitive objects of which we have knowledge (since we see them as they were five or more billion years ago). For various reasons there is a controversy over whether quasars are truly cosmological.

14. The cosmological constant must be very small, since no effect has been observed in the solar system. Einstein, who introduced the cosmological constant, came to be the leading disbeliever.

15. For positive curvature, the volume, and hence, the number of sources, increases less rapidly than r^3. Specifically, $V = 4/3\pi r^3(1-kr^2/5)$, where k is the curvature.

16. A Sandage, "The Redshift–Distance Relation II," *Astrophysical Journal* (1972), 178:1. The value given was $q = 0.96 \pm 0.4$.

17. J. R. Gott et al., "Will the Universe Expand Forever?" *Scientific American,* March 1976. The original paper was Gott et al., *Astrophysical Journal* (1974), 194:543. See also B. M. Tinsley, "The Cosmological Constant and Constant Change," *Physics Today* (June 1977).

18. Current theory says that only hydrogen (about 75%) and helium (over 23%), plus a small amount of deuterium and lithium were produced in the big bang. All other elements are manufactured through nucleosynthesis in stellar interiors.

19. Gott et al., "Expand Forever?"

20. *Science News* (May 10, 1980) and *Mosaic* (November–December 1980).

21. A "black body" is an idealized object used by physicists in describing radiation from a hot body. A black body is a perfect absorber and a perfect emitter. The total energy radiated by a black body is given by a constant times the area times the temperature raised to the fourth power (kAT^4). The spectrum of a black body (energy radiated versus wavelength) is a unique curve for a given temperature (see fig. 14.13 for a black body curve corresponding to 2.7K). Max Planck's derivation of the mathematical form of the black body curve at the turn of the century represents the birth of the quantum theory.

22. First published by R. A. Alpher and R. C. Herman, "On the Relative Abundance of the Elements," *Physical Review* (1948), 74:1737. Gamow was a student of Friedmann.

23. With photons. Using neutrinos we could look back much further, to a red shift z of about 10^{10} (a temperature of about 10^{10}K), when the universe was about 1 second old. See S. Weinberg, *The First Three Minutes* (New York: Basic Books, 1977).

24. J. R. Gott, *Nature* (1982), 245:304. Draws on work by A. Guth.

25. Fred Hoyle and J. V. Narlikar have even proposed that the universe is *not* expanding, that particle masses increase with time via Mach's principle (a given particle interacts with more and more matter, as a function of time.) The result is that the wavelengths associated with electronic transitions in atoms decrease with time. They were longer in the past, so that distant galaxies show "red shifts." A consequence of this model is that atoms, and thus everything else, shrink with time. The universe is not expanding, we are contracting! Eddington made a similar suggestion in 1932.

26. On the other hand, most of the entropy (a measure of "disorder" in the universe) is in the form of the 3K background. The entropy increase due to irreversible processes involving matter (including those which lead to living things) during the last 10^{10} years is totally negligible. See Weinberg, *The First Three Minutes* or Harrison, *Cosmology*.

27. D. Sciama, "Cosmology Before and After Quasars," *Scientific American* (September 1967).

28. One of the early excursions into this area was Brandon Carter, "Large Number Coincidences and the Anthropic Principle in Cosmology," in M. S. Longair, ed. *Confrontation of Cosmological Theories with Observational Data* (Dordrecht, Netherlands: Reidel, 1974).

29. E. R. Harrison, "Why the Sky is Dark at Night," Physics Today, February 1974. The problem is that since the volume of a shell of space (and hence the number of stars in it) increases with the square of its radius, but the light from each star (or galaxy) in the shell *decreases* with the distance squared, the two factors neatly cancel each other, so that each shell, whether near or far, contributes equally to the brightness of the sky as seen from the earth. There are infinitely many such shells, so that the sky, if the universe is infinite in space and time, and static, should be infinitely bright, or at least as bright as the surface of a star. This is "Olber's paradox."

30. The weak interaction is involved in "beta decay," as when a neutron decays into a proton (the reverse occurs in the first stage of the proton-proton cycle by which energy is generated in the sun. See chapter 15). The weak interaction can change the *flavor* of a quark. The strong interaction is what binds the nucleus together and ultimately derives from the exchange of colored "gluons" between quarks. The strong force can change the color of a quark.

31. The *Scientific American* reprint volume *Particles and Fields* (San Francisco: Freeman, 1980) provides an excellent introduction to these ideas. See also M. S. Turner and D. N. Schramm, "Cosmology and Elementary Particle Physics," *Physics Today* (Sept. 1979).

32. Steven Weinberg, *The First Three Minutes*.

33. "Grand Unification: An Elusive Grail," *Mosaic* (Sept.–Oct. 1979), p. 12.

34. *Ibid.*, p. 12. The photon, for example, is a boson (Bose-Einstein particle); the electron (or neutron or proton) is a Fermion (Fermi-Dirac particle). On the

nunification of gravity and the other forces of nature, see D. Z. Freedman and Peter von Nieuwenhuizen, "Supergravity and the Unification of the Laws of Physics," *Scientific American* (Feb. 1978).

35. "Grand Unification: An Elusive Grail," p. 12.

15. Gravitational Collapse

1. J. A. Wheeler, in M. Rees, R. Ruffini, and J. A. Wheeler, *Black Holes, Gravitational Waves, and Cosmology* (New York: Gordon and Breach, 1974), p. 307.

2. Especially James Hutton in the late eighteenth century, Charles Lyell in the nineteenth, and of course, Darwin and Wallace in the early to mid-nineteenth century.

3. C. F. v. Weizacker, *Physikalische Zeitschrift* (1937), 38:176; H. Bethe and C. L. Critchfield, *Physical Review* (1938), 54:248. See also M. Schwarzschild, *Structure and Evolution of the Stars* (Princeton: Princeton University Press, 1958), § 10.

4. Even the theory of energy production in the sun and other stars is not without its difficulties. In particular, the neutrino experiment of Davis has found only 20 percent of the neutrinos expected on the basis of the accepted theory. See J. N. Bahcall, "Neutrinos from the Sun," *Scientific American* (July 1969). Reprinted in *New Frontiers in Astronomy* (San Francisco: Freeman, 1970). A much more technical article is Bahcall, "Solar Neutrino Experiment," *Reviews of Modern Physics* (1978), 50:881. The experiment consists of detecting neutrinos in a tank of 100,000 gallons of perchloroethylene (cleaning fluid, C_2Cl_4) in the Homestake goldmine in South Dakota. If neutrinos have mass (see chapter 14), their "oscillations" from one state to another could explain the low neutrino yield. Alternatively, since neutrinos leave the sun without appreciable chance of scattering or absorption, whereas photons take on the order of 10^4 years to diffuse outward to the surface, our present neutrino flux may reflect a lowered central temperature in the sun that will not be manifested in its luminosity and surface temperature for up to 10,000 years. Everything we know of the structure of stars and how they evolve depends on this model of stellar evergy production (and very similar ones), so that it is most unlikely that the theory is in error.

5. If the luminosity is proportional to the third power of the mass, it means that when the mass is increased by a factor of 10, say, from the mass of the sun to 10 times that value, the luminosity or power output will increase, not by 10 times, but by 10^3 or 1000 times.

6. Such a star would be a red dwarf star like Barnard's star, the second nearest star to the sun and the star that shows the greatest annual motion ("proper motion") across the face of the sky. Barnard's star may also be the star near the sun most likely to have a planetary system of some sort.

7. The luminosity or brightness depends on the area of the star and on the fourth power of the temperature. Thus if we compare a red supergiant star like Betelgeuse, with a radius nearly 1000 times that of the sun but with a surfa temperature of only 3000 K, half that of the Sun, with Sirius B, the white dwarf companion of the brightest star in the sky, 100 times smaller than the sun but with a temperature nearly twice as high, we find the following: the area of

Betelgeuse is ten billion times that of the white dwarf; Sirius B at a temperature four times that of Betelgeuse emits $4^4 = 256$ times as much energy per square centimeter. Betelgeuse is thus 10 billion/256 = 40 million times brighter.

8. The curve dividing nuclei that release energy by fusion from those that release energy by fission peaks at about iron-57. Reactions involving heavier nuclei absorb energy; they are endoergic. This is precisely the distinction between thermonuclear weapons (or the hoped-for twenty-first-century thermonuclear reactor) and fission weapons (or the nuclear reactor).

9. There are other supernova scenarios, but we will not go into them.

10. See, for example, John C. Brandt, "Petrographs and Petroglyphs of the Southwest Indians," in K. Brecher and M. Fiertag, eds., *Astronomy of the Ancients,* Cambridge: MIT, 1979.

11. F. R. Stephenson and D. H. Clark, "Historical Supernovas," *Scientific American* (June 1976).

12. A name bestowed by John Wheeler.

13. One, the companion of the star 40 Eridani (known also as o₂ Eridani) is rather easily seen in small telescopes. This star is known as 40 Eridani B, and its nearby companion, 40 Eridani C, is a red dwarf star, making this triple star system an unusually interesting one for the observer with a small or medium-sized telescope.

14. Angular momentum conservation being assumed. In fact some angular momentum is undoubtedly lost, being transferred to the expanding envelope, but the argument is generally correct.

15. Whatever the nature of the disturbance that gave rise to the pulsed radiation, it could not propagate faster than light. Thus if the entire surface of the star were involved, a pulse with a period of 1/30 second implies a diameter no larger than $(1/30) \, (186,000) = 6200$ miles, a size typical of a white dwarf star. In fact, as we shall see, the source is much smaller.

16. The power output of the sun is 4×10^{33} ergs/sec or 4×10^{26} watts.

17. Named for the relativity theorist Karl Schwarzschild, a contemporary of Einstein. The space-time geometry appropriate to a nonrotating, uncharged black hole (outside the matter) was discovered by him.

18. J. A. Wheeler, "Our universe, the Known and the Unknown," *American Scientist* (1968), 56:1.

19. Mathematically there is a singularity; whether singularities are allowed in the real world remains to be seen. "Isn't the electron a singularity?" Eugene Wigner to one of the authors.

20. See, for example, S. Hawking and G. F. R. Ellis, *The Large Scale Structure of Spacetime* (Cambridge: Cambridge University Press, 1973).

21. The name of Szekeres is often associated with the transformation of ordinary space time that leads to this diagram. The first to use it in its modern form was David Finkelstein, who found that Eddington had employed it in 1924.

22. The reader who is enchanted by such ideas should read W. J. Kaufmann, *The Cosmic Frontiers of General Relativity* (Boston: Little Brown, 1977).

23. Especially as the less massive and therefore "younger" star approaches the red giant or supergiant stage and begins to swell.

24. Cygnus X-1, the first x-ray source found in Cygnus, was discovered by the Uhuru satellite. It may be the remnant of a supernova which occurred in 1408. See *Sky and Telescope,* October 1979, p. 323.

25. This may provide some help to the reader in understanding the seemingly paradoxical situation of the universe expanding from a very dense or even singular state.

26. Recall that the Schwarzschild radius is given by $R = 2GM/c^2$. The volume is $4/3\pi R^3$ so that the density $\rho = M/V = (\text{constant})/M^2$.

27. There is recent evidence of a galactic halo that could raise this figure to 10^{12} M (one trillion solar masses). Kepler's third law (chapter 9) applied to the galaxy, assuming the sun's distance from the center to be 30,000 light years and its period of revolution about the galactic center to be 250 million years, gives about 10^{11} M.

28. See S. Hawking, "The Quantum Mechanics of Black Holes," *Scientific American* (Jan. 1977). Reprinted in the reprint volume *Cosmology +1* (San Francisco: Freeman, 1977).

29. That is, the second law of thermodynamics. A good discussion can be found in P. C. W. Davies, *Space and Time in the Modern Universe* (Cambridge, England: Cambridge University Press, 1977). Harrison, in his *Cosmology*, gives a good discussion of the cosmological implications.

Epilogue

1. A. Einstein, *Ideas and Opinions,* S. Bergmann trans. (New York: Bonanza Books, 1954). p. 225. Other quotations in this paragraph are from the same brief address.

2. H. Morgenau, *The Nature of Physical Reality* (New York: McGraw-Hill, 1950).

3. B. S. DeWitt and N. Graham, eds., *The Many-Worlds Interpretation of Quantum Mechanics* (Princeton, N.J.: Princeton University Press, 1973).

4. B. S. DeWitt, "Quantum Mechanics and Reality," in *Physics Today* (Sept. 1970), p. 33.

5. Gerald Holton, *Thematic Origins of Scientific Thought* (Cambridge: Harvard University Press, 1973).

6. Einstein attributes the phrase "pre-established harmony" to Leibnitz, for whom it referred to the resolution of the Cartesian duality. The context of Einstein's usage clearly points to a broader Platonic or Pythagorean kind of harmony.

7. H. Frankfort, *Kingship and the Gods* (Chicago: University of Chicago Press, 1948), p. 66.

Bibliography

Overview

Brody, B. A. and N. Capaldi, eds. *Science: Men, Methods, Goals.* New York: Benjamin, 1968.

Brown, J. B., ed. *Exploring the Universe.* New York: Oxford, 1971.

Butterfield, H. *The Origins of Modern Science.* New York: Macmillan, 1958.

Christianson, G. E. *This Wild Abyss.* New York: Macmillan, 1978.

de Santillana, G. *Origins of Scientific Thought.* Chicago: University of Chicago Press, 1961.

de Solla Price, D. J. *Science Since Babylon.* New Haven: Yale University Press, 1961.

Dijksterhus, E. J. *The Mechanization of the World Picture.* Oxford: Clarendon, 1961.

Dreyer, J. L. E. *A History of Astronomy from Thales to Kepler.* New York: Dover, 1953. (Reprint of 1906 ed. entitled *A History of the Planetary Systems from Thales to Kepler.*)

Eiseley, L. *The Firmament of Time.* New York: Atheneum, 1970.

Feyerabend, P. *Against Method.* London: Humanities Press, 1975.

Feynman, R. *The Character of Physical Law.* Cambridge: MIT, 1967.

Gillispie, C. C. ed. *Dictionary of Scientific Biography.* 16 vols. New York: Scribners, 1970–80.

Grunbaum, A. *Philosophical Problems of Space and Time.* 2nd enlarged ed. Boston: Reidel, 1973.

Hall, A. R. *The Scientific Revolution, 1500–1800.* Boston: Beacon, 1956.

Hall, A. R. and M. B. Hall. *A Brief History of Science.* New York: New American Library, 1964.

Holton, G. *Thematic Origins of Scientific Thought.* Cambridge: Harvard University Press, 1973.

Jammer, M. *Concepts of Space.* Cambridge: Harvard University Press, 1957.

Koestler, A. *The Sleepwalkers.* New York: Grosset and Dunlap, 1959.

Kuhn, T. S. *The Copernican Revolution.* New York: Vintage, 1957.

——— *The Structure of Scientific Revolutions.* 2d enlarged ed. Chicago: University of Chicago Press.

Matson, F. W. *The Broken Image*. New York: Doubleday, 1966.

Munitz, M. K. *Space, Time, and Creation*. Glencoe, Ill. Free Press, 1957.

——— *Theories of the Universe*. Glencoe, Ill.: Free Press, 1962.

Pannekoek, A. *A History of Astronomy*. New York: Wiley, 1961.

Popper, K. *The Logic of Scientific Discovery*. New York: Harper and Row, 1965.

Reichenbach, H. *The Philosophy of Space and Time*. New York: Dover, 1957.

Roszak. T. *Where the Wasteland Ends*. New York: Doubleday, 1973.

Sambursky, S. *Physical Thought from the Presocratics to the Quantum Physicists*. New York: Pica, 1975.

Sarton, G. *A History of Science*. 2 vols. New York: Norton, 1959.

Snow, C. P. *The Two Cultures and a Second Look*. New York: New American Library, 1963.

Taton, R. *A General History of the Sciences*. 4 vols. London: Thames and Hudson, 1963–66.

Toulmin, S. *The Philosophy of Science: An Introduction*. New York: Harper, 1960.

Toulmin, S. and J. Goodfield. *The Fabric of the Heavens*. New York: Harper and Row, 1961.

——— *The Discovery of Time*. New York: Harper and Row, 1965.

Wigner, E. P. *Symmetries and Reflections*. Cambridge: MIT, 1970.

Prehistoric and Early Historic Periods

Atkinson, R. J. C. *Stonehenge*. London: H. Hamilton, 1956.

Aveni, A. F. ed. *Archaeoastronomy in Pre-Columbian America*. Austin: University of Texas Press, 1975.

——— *Native American Astronomy*. Austin: University of Texas Press, 1977.

——— *Skywatchers of Ancient Mexico*. Austin: University of Texas Press, 1980.

——— "Venus and the Maya." *American Scientist* (1979), May–June.

Barnard, M. *The Mythmakers*. Athens: Ohio University Press, 1967.

Blacker, C. and M. Loewe, eds. *Ancient Cosmologies*. London: Allen and Unwin, 1975.

Brecher, K. and M. Freitag, eds., *Astronomy of the Ancients*. Cambridge: MIT, 1979.

Budge, E. A. W. *The Egyptian Book of the Dead*. New York: Dover, 1967.

Burl, A. *The Stone Circles of the British Isles*. New Haven: Yale University Press, 1976.

Campbell, J. *The Hero with a Thousand Faces*. New York: Pantheon, 1949.

Clark, R. T. R. *Myth and Symbol in Ancient Egypt*. New York: Grove, 1960.

Eliade, M. *Cosmos and History*. W. R. Trask, trans. New York: Harper, 1959.

——— *Gods, Goddesses, and Myths of Creation*. New York: Harper, 1967.

Frankfort, H., H. A. Frankfort, J. A. Wilson, and T. Jacobson. *Before Philosophy*. Baltimore: Penguin, 1946.

Haddingham, E. *Circles and Standing Stones*. New York: Doubleday Anchor, 1976.

Hawkins, G. *Beyond Stonehenge*. New York: Harper, 1973.

Hawkins, G. and J. B. White. *Stonehenge Decoded*. New York: Dell, 1965.

Heidel, A. *The Babylonian Genesis*. Chicago: University of Chicago Press, 1951.

Hodson, F. R. ed. *The Place of Astronomy in the Ancient World*. London: Oxford, 1974.

Kramer, S. N. *Mythologies of the Ancient World*. New York: Doubleday Anchor, 1961.

Krupp, E. C. ed. *In Search of Ancient Astronomies*. New York: Doubleday, 1977.

Needham, J. *Science and Civilization in China*. Vol. 3. Cambridge: Cambridge University Press, 1954.

Newham, C. A. *The Astronomical Significance of Stonehenge*. Leeds: John Blackburn, 1972.

Pritchard, J. B. ed. *Ancient Near Eastern Texts Relating to the Old Testament*. Vol. 1. Princeton: Princeton University Press, 1958.

de Santillana, G. and H. von Dechend. *Hamlet's Mill*. Boston: Gambit, 1969.

Teeple, J. E. "Maya Astronomy." In *Contributions to American Archaeology*. Washington, D.C.: Carnegie Institute, 1931.

Thom, A. *Megalithic Sites in Britain*. Oxford: Clarendon, 1967.

—— *Megalithic Lunar Observatories*. Oxford: Clarendon, 1971.

—— *Megalithic Remains in Britain and Brittany*. Oxford: Clarendon, 1978.

van der Waerden. B. L. *Science Awakening*. 2 vols. Leyden: Noordhoff, 1974–75.

Antiquity

Aristotle. *On the Heavens (De Caelo)*. W. K. C. Guthrie, trans. London: Heinnemann, 1934.

Burnet, J. *Greek Philosophy. Part I, Thales to Plato:* London: Macmillan, 1914.

Clagett, M. *Greek Science in Antiquity*. New York: Abelard-Schuman, 1955.

Cohen, M. and I. E. Drabkin. *A Source Book in Greek Science*. New York: McGraw-Hill, 1948.

Cornford, F. M. *Plato's Cosmology: The Timaeus of Plato*. New York: Humanities Press, 1952.

Dicks, D. R. *Early Greek Astronomy to Aristotle*. Ithaca: Cornell University Press, 1970.

Dreyer, J. L. E. *A History of Astronomy from Thales to Kepler*. New York: Dover, 1953.

Duhem, P. *The Aim and Structure of Physical Theory*. P. P. Wiener, trans. New York: Atheneum, 1981.

Farrington, B. *Greek Science*. London: Penguin, 1953.

Freeman, K. *An Ancilla to the Pre-Socratic Philosophers*. Cambridge: Harvard University Press, 1957.

Gershenson, D. and D. Greenberg. *Anaxagoras and the Birth of the Scientific Method*. New York: Blaisdell, 1964.

Guthrie, W. K. C. *A History of Greek Philosophy. Vol 2. The Presocratic Tradition from Parmenides to Democritus.* Cambridge: Cambridge University Press, 1965.

Heath, T. L. *Aristarchus of Samos.* Oxford: Clarendon, 1913.

────── *Greek Astronomy.* New York: AMS, 1969. (Reprint of 1932 ed.)

Kahn, C. H. *Anaximander and the Origins of Greek Cosmology.* New York: Columbia University Press, 1952.

Kirk, G. S. and J. E. Raven. *The Presocratic Philosophers.* Cambridge: Cambridge University Press, 1957.

McKeon, R. P. *The Basic Works of Aristotle.* New York: Random House, 1941.

Neugebauer, O. *The Exact Sciences in Antiquity.* New York: Dover, 1969.

────── *A History of Ancient Mathematical Astronomy.* 3 vols. New York: Springer-Verlag, 1975.

Pederson, O. *A Survey of the Almagest.* Odense: Odense, 1974.

Ptolemy. *Mathematical Syntaxis,* or *The Almagest.* In R. C. Taliafero, trans. *Great Books of the Western World.* Vol 16. Chicago: Encyclopedia Britannica, 1952.

Sambursky, S. *The Physical World of the Greeks.* M. Dagut, trans. London: Routledge, 1956.

────── *The Physical World of Late Antiquity.* New York: Basic Books, 1962.

Vlastos, G. *Plato's Universe.* Seattle: University of Washington Press, 1975.

The Middle Ages

Bacon, Roger. *Opus Maius.* R. B. Burke, trans. Philadelphia: University of Pennsylvania Press, 1928.

Titus Lucretius Carus. *On the Nature of Things.* W. H. D. Rouse, trans. Cambridge: Harvard University Press, 1947.

Crombie, A. C. *Robert Grosseteste and the Origins of Experimental Science.* Oxford: Clarendon, 1953.

────── *Augustine to Galileo: The History of Science. A.D. 400–1600.* Cambridge: Harvard, 1961.

Dales, Richard C. *The Scientific Achievement of the Middle Ages.* Philadelphia: University of Pennsylvania Press, 1973.

Dijksterhus, E. J. *The Mechanization of the World Picture.* Oxford: Clarendon, 1961.

Duhem, P. *To Save the Appearances.* Chicago: University of Chicago Press, 1969.

Easton, C. S. *Roger Bacon and His Search for a Universal Science.* Oxford: Blackwell, 1952.

Grant, E. *Physical Science in the Middle Ages.* New York: Wiley, 1971.

────── *A Source Book in Medieval Science.* Cambridge: Harvard, 1974.

Graubard, M. *Motivations, Tools and Theories of Pre-Modern Science.* Minneapolis: Burgess, 1967.

Haskins, C. H. *The Renaissance of the 12th Century.* Cambridge: Harvard University Press, 1927.

Koyre, A. *From the Closed World to Infinite Universe*. Baltimore: Johns Hopkins University Press, 1957.

Palter, R. M. ed. *Toward Modern Science*. Vol. 1. New York: Noonday Press, 1961.

Shapiro, H. *Motion, Time, and Place According to William Ockham*. St. Bonaventure, New York: Franciscan Institute, 1957.

Stahl, W. H. *Roman Science*. Madison: University of Wisconsin Press, 1962.

Thompson, A. H. *Bede, His Life, Times, and Writings*. Oxford: Clarendon, 1969.

Wilson, C. *William Heytesbury*. Madison: University of Wisconsin Press, 1956.

Thorndike, L. *A History of Magic and Experimental Science*. 8 vols. New York: Macmillan and Columbia University Press, 1923–58.

Copernicus, Kepler, Galileo

Armitage, A. *Copernicus*. New York: T. Yoseloff, 1957.

Baumgardt, C. *Johannes Kepler: His Life and Letters*. New York: Philosophical Library, 1951.

Beer, A. and P. Beer. *Kepler. Vistas in Astronomy* (1975) vol. 18.

Beer, A. and K. Strand. *Copernicus. Vistas in Astronomy* (1975) vol. 17.

Burtt, E. A. *Metaphysical Foundations of Modern Science*, rev. ed. New York: Doubleday, 1954.

Caspar, M. *Kepler*. C. D. Hellman, trans. New York: Abelard-Schuman, 1959.

Copernicus. *De Revolutionibus Orbium Celestium*, 1543. (There are 3 English translations. The first appeared in vol. 16 of *Great Books of the Western World* [Chicago: Encyclopedia Britannica]. This 1952 translation is by C. G. Wallis. In 1976 A. M. Duncan's translation was published by Barnes and Noble; in 1978, E. Rosen's translation was published by Johns Hopkins University Press.)

Cusa, Nicholas of. *Of Learned Ignorance*. G. Heron, trans. London: Routledge, 1954.

Drake, S. *Discoveries and Opinions of Galileo*. New York: Doubleday Anchor, 1957.

────── *Galileo Studies*. Ann Arbor: University of Michigan, 1970.

────── *Galileo at Work*. Chicago: University of Chicago Press, 1978.

Dreyer, J. L. E. *Tycho Brahe*. New York: Dover, 1963. Rept. of 1890 ed.

Dugas, R. *Mechanics in the Seventeenth Century*. F. Jacquot, trans. New York: Central Book Co., 1958.

────── *A History of Mechanics*. J. R. Maddox, trans. New York: Central Book Co., 1955.

Gade, J. A. *The Life and Times of Tycho Brahe*. Princeton: Princeton University Press, 1947.

Galileo. *Dialogue Concerning the Two Chief World Systems*. S. Drake, trans. Berkeley: University of California Press, 1967.

────── *Discourse Concerning Two New Sciences*. H. Crew and A. deSalvio, trans. New York: Dover, 1954.

―――― *The Sidereal Messenger.* E. S. Carlos, trans. London: Dawson's of Pall Mall, 1959. Rept. of 1890 ed.

Geymonat, L., *Galileo Galilei,* S. Drake, trans. New York: McGraw-Hill, 1965).

Holton, G. "Johannes Kepler's Universe, Its Physics and Metaphysics." *American Journal of Physics* (1956), 24:340.

Kepler, J. *Epitome of Copernican Astronomy,* Books 4 and 5, and *The Harmonies of the World* (Harmonices Mundi) Book 5. In *Great Books of the Western World,* vol. 16. Chicago: Encyclopedia Britannica, 1952.

Koyre, A. *From the Closed World to the Infinite Universe.* Baltimore: Johns Hopkins University Press, 1957.

Kuhn, T. S. *The Copernican Revolution.* New York: Vintage, 1957.

Lovejoy, A. *The Great Chain of Being.* Cambridge: Harvard University Press.

McMullin, E. *Galileo, Man of Science.* New York: Basic Books, 1968.

Ronan, C. *Galileo.* New York: Putnam, 1974.

Rosen, E. Three *Copernican Treatises.* New York: Columbia University Press, 1939. Contains the *Commentariolus,* the *Narratio Prima of Rheticus,* and the *Letter Against Werner.*

de Santillana, G. *The Crime of Galileo.* Chicago: University of Chicago Press, 1955.

Shea, W. R. *Galileo's Intellectual Revolution.* New York: Science History, 1977.

Singer, D. *Giordano Bruno: His Life and Thought.* New York: Abelard-Schuman, 1950.

Small, R. *An Account of the Astronomical Discoveries of Kepler.* Rpt. of 1804 ed. Madison: University of Wisconsin Press, 1963.

Newton

Armitage, A. *Edmund Halley.* London: Nelson, 1966.

Ault, D. D. *Visionary Physics: Blake's Response to Newton.* Chicago: Chicago University Press, 1975.

Ball. W. W. R. *Essay on Newton's Principia.* New York: Johnson Reprint Co., 1972. Rept. of 1893 ed.

Butterfield, H. *The Origins of Modern Science.* New York: Macmillan, 1958.

Cohen, I. B. *The Newtonian Revolution.* Cambridge: Cambridge University Press, 1980.

―――― *Introduction to Newton's Principia.* Cambridge: Harvard University Press, 1971.

Cohen, I. B. and A. Koyre. *Isaac Newton's Philosophae Naturalis Principia Mathematica, The Third Edition with Variant Readings.* Cambridge: Harvard University Press, 1972.

Frye, N. *Fearful Symmetry.* Princeton: Princeton University Press, 1947.

Gillispie, C. C. *The Edge of Objectivity.* Princeton: Princeton University Press, 1960.

Hall, A. R. *Philosophers at War: The Quarrel Between Newton and Leibniz.* New York: Cambridge University Press, 1980.

Hall, A. R. *The Rise of Modern Science*. New York: Harper, 1962.

Herivel, J. *The Background to Newton's Principia: A Study of Newton's Dynamical Research in the years 1664–84*. Oxford: Clarendon, 1965.

Koyre, A. *Newtonian Studies,* Chicago: University of Chicago Press, 1965.

Manuel, F. *A Portrait of Isaac Newton*. Washington: New Republic Press, 1979.

More, L. T. *Isaac Newton*. New York: Dover, 1962.

Nicholson, M. *Science and Imagination*. Ithaca: Great Seal Books (Cornell), 1956.

—————— *The Breaking of the Circle: Studies in the Effect of the New Science on Seventeenth Century Poetry,* Evanston, Ill.: Northwestern University Press, 1949.

—————— *Newton Demands the Muse*. Princeton: Princeton University Press, 1946.

Newton, I. *Mathematical Principles of Natural Philosophy*. A. Motte, trans. rev. F. Cajori. Berkeley: University of California Press, 1934.

Newton, I. *Opticks* (Fourth Edition, 1730). New York: Dover, 1952.

Tilyard. E. M. W. *The Elizabethan World Picture*. New York: Random House, 1959.

Turnbull, W. H. ed. *Isaac Newton, The Correspondence*. Cambridge: Cambridge University Press, 1959.

Westfall, R. S. *Never at Rest*. Cambridge: Cambridge University Press, 1981. As of the date of its publication, the definitive Newton biography.

Whiteside, D. *The Mathematical Papers of Isaac Newton*. Cambridge: Cambridge University Press, 1967.

Einstein

Aichelburg, P. C. and Roman U. Sexl eds. *Albert Einstein, His Influence on Physics, Philosophy and Politics*. Wiesbaden: Friedrich Vieweg and Son, 1979.

Bernstein, J. *Einstein*. New York: Fontana, 1973.

Clark, R. W. *Einstein, the Life and Times*. New York: World Publishers, 1971.

Einstein, A. *Relativity*. New York: Crown, 1961.

Einstein, A. "Autobiographical Notes." In P. A. Schilpp, ed. *Albert Einstein: Philosopher Scientist*. New York: Harper and Row, 1959.

Frank, P. *Einstein, His Life and Times*. New York: Knopf, 1967.

French, A. P. ed. *Einstein: A Centenary Volume*. Cambridge: Harvard University Press, 1979.

Hoffman, B. *Albert Einstein, Creator and Rebel*. New York: New American Library, 1972.

Hoffman, B. and H. Dukas, eds. *Albert Einstein, the Human Side*. Princeton: Princeton University Press, 1979.

Miller, A. I. *Albert Einstein's Special Theory of Relativity*. Reading, Pa.: Addison-Wesley, 1981.

Swenson, L. S. Jr. *Genesis of Relativity*. New York: Burt Franklin, 1979.

Taylor, E. F. and J. A. Wheeler. *Space-Time Physics*. San Francisco: Freeman, 1966.

Whittaker, E. T. *A History of the Theories of Aether and Electricity*. 2 vols. New York: Harper Torchbooks, 1960.

Woolf, H. ed. *Some Strangeness in the Proportion*. Reading, Pa.: Addison-Wesley, 1980.

Relativity

Bergmann, P. G. *The Riddle of Gravitation*. New York: Scribners, 1968.

Berry, M. *Principles of Cosmology and Gravitation*. Cambridge; Cambridge University Press, 1976.

Brehme, R. W. "Curved Space and Gravitation." *American Journal of Physics* (1965) 33:383, 713.

Callahan, J. J. "The Curvature of Space in a Finite Universe." *Scientific American,* August 1976.

Einstein, A. *Relativity*. New York: Crown, 1961.

Davies, P. C. W. *Space and Time in the Modern Universe*. London: Cambridge University Press, 1977.

Grunbaum, A. *Philosophical Problems of Space and Time*. 2d enlarged ed. Dordrecht: Reidel, 1973.

Jammer, M. *Concepts of Space*. Cambridge: Harvard University Press, 1954.

Kaufmann, W. J. *General Relativity and Cosomology*. New York: Harper and Row, 1977.

———— *The Cosmic Frontiers of General Relativity*. Boston: Little, Brown, 1977.

Lanczos, C. *Albert Einstein and the Cosmic World Order*. New York: Wiley, 1965.

Mehra, J. *Einstein, Hilbert, and the Theory of Gravitation*. Dordrecht: Reidel, 1974.

Misner, C. W., K. S. Thorne, and J. A. Wheeler. *Gravitation*. San Francisco: Freeman, 1973.

Munitz, M. K. *Space, Time, and Creation*. Glencoe, Ill. Free Press, 1957.

Reichenbach, H. *Philosophy of Space and Time*. New York: Dover, 1957.

Sciama, D. W. *The Unity of the Universe*. New York: Doubleday, 1961.

———— "Inertia." *Scientific American,* February 1957.

Tauber, G. ed. *Albert Einstein's Theory of General Relativity*. New York: Crown, 1979.

Will, C. M. "Gravitation Theory." *Scientific American,* November 1974.

Cosmology

Berry, M. *Principles of Cosmology and Gravitation*. Cambridge: Cambridge University Press, 1976.

Bondi, H. *Cosmology*. Cambridge: Cambridge University Press, 1952.

Calder, N. *Violent Universe*. New York: Viking, 1970.

Harrison, E. R. *Cosmology*. Cambridge: Cambridge University Press, 1981.

Hoyle, F. *Astronomy and Cosmology*. San Francisco: Freeman, 1975.

Kaufmann, W. J. *The Cosmic Frontiers of General Relativity*. Boston: Little, Brown, 1977.

Misner, C. W. K. S. Thorne, and J. A. Wheeler. *Gravitation*. San Francisco: Freeman, 1973.

Lorentz, H. A., A. Einstein, H. Minkowski, and H. Weyl. *The Principle of Relativity*. New York: Dover, 1923.

Munitz, M. K. *Space, Time, and Creation*. Glencoe, Ill.: Free Press, 1957.

North, J. D. *The Measure of the Universe*. Oxford: Clarendon, 1956.

Rowan-Robinson, M. *Cosmology*. Oxford: Clarendon, 1977.

Sexl, R. and H. Sexl. *White Dwarfs—Black Holes*. New York: Academic Press, 1979.

Silk, J. *The Big Bang*. San Francisco: Freeman, 1980.

Singh, J. *Great Ideas and Theories of Modern Cosmology*, 2d. ed. New York: Dover, 1970.

Weinberg, S. *The First Three Minutes*. New York: Basic Books, 1977.

Black Holes

Hawking, S. J. "The Quantum Mechanics of Black Holes." *Scientific American*, January 1977.

Pasachoff, J. *Contemporary Astronomy*. 2d ed. Philadelphia: Saunders, 1981.

Schwarzschild, M. *Structure and Evolution of the Stars*. Princeton: Princeton University Press, 1958.

Shipman, H. L. *Black Holes and the Infinite Universe*. Boston: Houghton Mifflin, 1976.

Taylor, J. *Black Holes: The End of the Universe*. New York: Random House, 1923.

Thorne, K. "The Search for Black Holes." *Scientific American*, December 1974.

Index